E-governance

Also by Perri 6

6 P, Seltzer K, Leat D and Stoker G
TOWARD HOLISTIC GOVERNANCE: The New Agenda in Government Reform (2002)

6 P with Jupp B
DIVIDED BY INFORMATION? The 'Digital Divide' that Really Matters and the Implications of the New Meritocracy (2001)

6 P, Seltzer K, Leat D and Stoker G
GOVERNING IN THE ROUND: Strategies for Holistic Government (1999)

6 P
THE FUTURE OF PRIVACY, vols I and II (1998)

6 P
HOLISTIC GOVERNMENT (1997)

6 P
ESCAPING POVERTY: From Safety Nets to Networks of Opportunity (1997)

6 P and Kendall J, eds
THE CONTRACT CULTURE IN PUBLIC SERVICES: Studies from the UK, Europe and the USA (1997)

6 P with Briscoe I
ON THE CARDS: Privacy, Identity and Trust in the Age of Smart Technologies (1996)

6 P with Jupp B and Bentley T
OPEN WIDE: Futures for Dentistry in 2010 (1996)

6 P and Randon A
LIBERTY, CHARITY AND POLITICS: Non-profit Law and Freedom of Speech (1995)

6 P
BIG IS BEAUTIFUL: Bringing East Central Europe into the European Union (1995)

6 P and Vidal I, eds
DELIVERING WELFARE: Repositioning Non-profit and Co-operative Action in Western European Welfare States (1994)

E-governance

Styles of Political Judgment
in the Information Age Polity

Dr Perri 6
Senior Research Fellow
Health Services Management Centre
University of Birmingham

First published 2004 by
PALGRAVE MACMILLAN
Houndmills, Basingstoke, Hampshire RG21 6XS and
175 Fifth Avenue, New York, N.Y. 10010
Companies and representatives throughout the world

PALGRAVE MACMILLAN is the global academic imprint of the Palgrave Macmillan division of St. Martin's Press, LLC and of Palgrave Macmillan Ltd. Macmillan® is a registered trademark in the United States, United Kingdom and other countries. Palgrave is a registered trademark in the European Union and other countries.

ISBN 978-1-349-51110-5   ISBN 978-0-230-00089-6 (eBook)
DOI 10.1057/9780230000896

This book is printed on paper suitable for recycling and made from fully managed and sustained forest sources.

A catalogue record for this book is available from the British Library.

Library of Congress Cataloging-in-Publication Data
6, Perri, 1960–
      E-governance : styles of political judgment in the information age
   polity / Perri 6.
         p. cm.
      Includes bibliographical references and index.

      1. Internet in public administration. 2. Internet in public administration –
   Case studies. 3. Political planning – Data processing. 4. Political planning –
   Data processing – Case studies. I. Title.

JF1525.A8A14 2004
352.3'8'0285—dc22   2003065842

10 9 8 7 6 5 4 3 2 1
13 12 11 10 09 08 07 06 05 04

Transferred to digital printing 2005

Contents

v

List of Boxes and Figures

Boxes

Figures

List of Abbreviations Used in the Text

CRM   Customer relations management, a management practice for handling all accounts and contacts with clients, supported by dedicated software

CSCA   computer supported collaborative argumentation, an analytical discipline devoted to the study of group decision making using digital tools

DoE   Department of Energy, a division of US federal government

DSS   Decision support system, any digitally supported system for providing capabilities for decision makers to access and make use of information

EDI   Electronic document interchange, a system of digitally supported filing and exchange of documents, for example, for carrying out transactions, now largely superseded by web-based e-procurement systems

EIA   Energy Information Administration

EPA   Environmental Protection Agency, an agency of US federal government

GDSC   Group decision support centre, a physical space in which digital resources are provided for meeting management for CSCA

GDSS   Group decision support system, a DSS designed for collaborative work rather than individual use

GIS   Geographical information system, a form of software in which hyperlinked information is graphically represented using representations of geographical distributions, which may be used as or linked to DSS

GSI   Government Secure Intranet – a UK system for ministerial departments and selected non-departmental public bodies

IPCC   Intergovernmental Panel on Climate Change, a UN sponsored scientific body charged with synthesis of available data on global warming

NEMS   National Energy Modelling System, a model run by the EIA within DoE to support policy analysis across US federal government

Acknowledgements

Cisco Systems plc (UK) generously provided funding for this project. I am grateful to Nick Penston, Business Development Manager for the Public Sector at Cisco, in particular, for early discussions on the research question, arranging the financial support and for invaluable suggestions and support throughout the project, and for his patience. I am grateful to the following people who were members of an advisory committee for the project: Matthew Taylor, Director, Institute for Public Policy Research; Professor Helen Margetts, School of Public Policy, University College, London; Bill Cook, Consultant, Public Sector Division, Ernst and Young; Margaret Moran MP, who was then Parliamentary Private Secretary to the Chancellor of the Duchy of Lancaster; Francis Bottomley, Business Development Manager, EDS UK; Sarah Tanburn, then Director of Regeneration, Brighton and Hove Council; Mark Gladwyn, then of Central Information Technology Unit (now part of the E-envoy's office), and with the Home Office.

The following people generously made themselves available to be interviewed, and also furnished me with papers and other materials: Tim Morgan and Martin Brader, then of ICL; Professor Clive Holtham, City University Business School. I am grateful to Gerry Stoker, Professor of Politics at the University of Manchester for his encouragement, help with contacts and for his advice, to Ian Christie then of the Local Futures Group for valuable information, to John Williams then of the New Local Government Network for pressing me to think through the implications of my argument for the scrutiny function, to Andy O'Brien of the Improvement and Development Agency for Local Government for valuable contacts, to Paula Tiihonnen, advisor and researcher to the Finnish Parliament's Committee for the Future for sending me materials otherwise unavailable to me.

Ian Kearns, Sharon Jones, Paul Waller, James Tansey, Paula Tiihonen, Martin Surtees Putterill, David Bloor, Don Kettl all provided comments on sections of an earlier draft text. Fred Thompson sent messages of confidence, encouragement and advice that always seemed to come at times when my own faith in this project faltered. Some of the material in Chapters 2, 3 and 5 is taken from an earlier unpublished paper, on which the following people gave me invaluable comments: Christine Bellamy, Paul Frissen, Brian Hogwood, Helen Margetts, Lawrence Pratchett, Fred

Thompson and Steve Woolgar. Sandy Borins, Charles Raab and Christine Bellamy have commented on other chapters in draft form.

Having assured them of confidentiality, I am unable to name the many people who were so generous with their time, working in the three case study organisations discussed in Chapter 6. However, I am deeply grateful to them all for being willing to be interviewed at such length, and for providing me with additional documents and information.

The visit to the US to conduct these interviews would not have been possible without the financial support from Willamette University, Salem Oregon, and the University of Toronto, specially arranged for me by Fred Thompson and Sandy Borins respectively.

I am also especially grateful to Dave and Lori Garrison for their unstinting hospitality in Washington DC during the period when I was conducting the case studies in the US.

I am very grateful to Deborah Davidson for handling the financial management arrangements at King's College London, and to Edward Peck for his continuing support.

I am also grateful to Wolfgang Michalski and Riel Miller of the Organisation for Economic Cooperation and Development, who gave me an opportunity to air some of these arguments in a paper for their conference 'Forum for the future' conference in March 2000 (6, 2001c). I am also grateful to Paul Frissen, Monica den Boer and Wim van de Donk of the Centre for Informatisation, Law and Public Administration at the Catholic University of Brabant for the invitation to give a keynote address on some of these arguments at their conference, in Tilburg University in May 2000 (6, 2001d). Professor Helen Margetts invited me to give a paper at the Political Studies Association conference at the LSE in April 2000 (6, 2001e). I am grateful to Helen Margetts, Stephen Coleman and Gita Subramanyam and other participants whose names I do not know for their comments on that occasion. Some of the argument in Chapters 8 and 9 was first aired in a short article on evidence-policy making (6, 2002a).

Steven Kennedy, Alison Howson, Beverley Tarquini and Kerry Coutts at Palgrave have been as patient and helpful publishers as an author could hope for.

None of these people's generosity with their time and advice should be taken as committing them to agreeing with my arguments, still less should they be held responsible for my errors.

1
Introduction

What is e-governance?

If this is indeed the information age, or the network society (Castells, 1996), and if this means that digital information systems are transforming organisations and daily life, then we should expect our systems of governance to undergo a similar upheaval. And so they are, but not necessarily in the obvious ways. This book is an attempt to explore both the threats and the opportunities presented by new information and knowledge technologies for governance and policy making, and to set out some specific challenges to politicians and senior policy executives in national and local government.

Yet this is not yet another study of electronic government. There has been over the last ten or twelve years a vast flood of reports, recommendations, books, enquiries, commissions and online debates about how to put government information on the World Wide Web, how to prepare public services for digital television and the mobile telephone, how to offer public services through these media, how and whether to run elections online, the importance of an infrastructure to support citizens in using these services safely and securely through everything from data protection regulation to digital signatures on smart cards, how parliaments can make themselves more accountable through the web, and how legislators can vote electronically, and many other similar problems. This book is about something else, something at least as important, and something which has been astonishingly neglected in all this vast recent literature.

For government is not just – or even principally – about the provision of public services, nor is it only about the process by which votes are cast or the opinions of individuals, pressure groups and businesses are delivered to politicians. What has been missing is any attempt to think through the business of *governance*, by which I mean here the analysis and understanding of problems by elected and appointed politicians and their senior staff, the making of policy, the process of deliberation and cogitation, the process of exercising and cultivating political judgment, the making of decisions, and the oversight and scrutiny that other politicians and regulators exercise. This is not necessarily the most fashionable use of the word 'governance', but it is one of the classical and canonical uses, to describe the exercise of public power for steering social systems.[1]

'Political judgment', in this book, means the organisational processes of coming to sense-making, understanding appreciation and decision about courses of action and about institutions in organisations with public authority accountable to senior elected officials and through them and through the media to the voting and taxpaying public (as opposed to, for example, settlement of popular attitudes, or the cognitive cunning or luck of individual leaders).

Yet these are the basic activities that we as citizens elect our politicians to conduct, and pay taxes to support the core civil service and the heart of local government to carry out. Indeed, these were the principal activities of government long before it became democratic or extended itself into the business of providing civilian services at all, and they remain at the heart of the task of governance. E-governance, in the sense of using digital technologies to support political judgment, has been too little studied (Kraemer and Dedrick, 1997), and yet it surely will be of fundamental importance to the nature of democratic life.

The silence on governance in the e government debate is all the more surprising because a generation ago and before the advent of the personal computer as an everyday household or briefcase item, back in the days of mainframe computers and now forgotten programming languages, the main area of discussion about the implications of the information revolution for government was concerned with precisely this. Some of the earliest applications of the first computers were in developing simple models of the US economy with which to advise the US Treasury. By the 1950s, the new generation of analysts of administration had begun work on the role of artificial intelligence in supporting decision making, in the hope of enhancing the rationality of political decision making (Simon, 1992a [1964], 1992b [1966], 1997 [1945] esp. ch. VIII, 1977 [1960]). By the mid-1970s, high hopes were being attached to the application of computer power to such techniques for policy analysis as linear programming (Greenberger *et al.*, 1976;

van Gunsteren, 1976; Stokey and Zeckhauser, 1978). Although the early romantic rationalism of those decades about the possible consequences for policy has evaporated, many new technologies have since been developed with which to support the work of governance, and these have begun to diffuse among policy makers throughout the world, and their use will surely have important consequences.

For example, during the 1980s, the first expert systems were deployed in governance, and formal models began to be experimented with, comprising hundreds and later thousands of equations in order to model and understand the possible impact of policy changes upon national economies, epidemiological patterns of the virulence of disease, and other problem structures. In the second half of that decade, spreadsheet software for personal computers became widely available and cheap, enabling much wider and faster access to the basic data for budget setting for those politicians who wanted to get involved in detail, rather than leaving it to finance directors and Treasury civil servants. Although these phenomena have been missing from all the sound and fury about the Internet and online communications, the 1990s and early 2000s were quietly a decade of very extensive innovation in the fields of group decision support systems, meeting management systems, decision analysis tools, creativity and lateral thinking tools, electronic whiteboard packages, electronic brainstorming support systems, decision conferencing systems, knowledge management systems, organisational memory capture systems, problem structuring tools for such approaches as soft systems methodology, and indeed groupware of every kind. Moreover, in that decade, approaches that are still loosely grouped together under the rubric of 'artificial intelligence' started to yield practical tools such as neural nets for decision modelling, digital agents, sophisticated pattern recognition systems, and new forms of neural net-based inductive modelling tools.

Hitherto, policy makers in countries such as Britain may seem to have been laggards in the take-up of these kinds of tools. By comparison, their opposite numbers in north America and Australasia, some parts of Scandinavia and in the Netherlands seem to have been much keener. It is for this reason that some of the case studies to which this book devotes the greatest attention are from the United States. As a general principle, of course, caution in taking up new technologies is not necessarily unwise. There is much to be said for allowing others to be the guinea pigs, to experiment and waste time on tools that turn out to be of little value, and to pick and choose from a more mature market. However, for reasons that will become clear in the course of this book, learning from others about 'what works', in this field, is not straightforward, because the situations, requirements and capabilities of different groups of ministers, professional civil service policy analysts, local government chief officers, regulators, boards, scrutiny committee members, can differ substantially and in ways that are not always obvious from an examination of their roles and formal powers. For the consequences of using such systems depend very heavily and in ways that are far from obvious upon the character of the policy-making group, and much more than they depend upon the technical features of the software.

Nevertheless, it seems highly likely that during the next few years, these kinds of systems will be bought and installed to support the work of policy makers throughout political systems. There are some indications that the use of such tools is something to which, for example, the British government is beginning to attach higher priority (Performance and Innovation Unit, 2000a). This flows in part from the rising importance of policy coherence, given the imperative for more 'joining up' horizontally at the level of public services and of regulation. In part it flows from the greater role for local authorities in policy making envisaged in recent legislation and current policy. In some part, too, it derives from the changing patterns of working among senior people coming into executive roles from senior management positions in the private sector, where much greater use is made than has historically been the case in the public sector, of digital tools to support decision making. In some part, it has to do with the greater professionalisation of the leadership cadre within political parties, and their growing commitment to the use of tools that will enable them not only to feel more professional, but to be able to demonstrate to their electorates that their political and policy judgments are not just matters of instinct but have some method or procedure behind them.

In the post-war decades of mainframe computing and right up to the late 1980s, most e-governance tools offered by the information systems industry were presented as ways of substituting for judgment, of eliminating ambiguity and uncertainty, or else simply as providing yet more information. These, understandably, had limited appeal to policy makers who know the inevitability of judgment, however precise the available data might be, and who know the importance of ambiguity in certain situations, and whose overriding need is not more but less information of greater value. However, more recently, tools have been both developed and presented in ways that support the manner in which policy makers can and must work: for example, many tools are now explicitly designed to support the subjective aspects of judgment.

Finally, as the price of the hardware and the software required falls to levels that can be afforded within the conventional information support budgets of most civil service departments, the argument of disproportionate expense for the expected gain has a declining weight. Therefore, it is particularly important to begin thinking now about how best to proceed.

Of course, it is too early to write any definitive assessment of what the consequences of the take-up and use by policy makers of these kinds of tools, jointly and severally, will be. However, some important things can be said now about the conditions for their effective use, what might be sensible goals and expectations for their use, some of the dangers of their use in particular ways. Some over-inflated expectations can usefully be challenged even now, but so too can some of the ways in which these tools are dismissed by many practising politicians and by many commentators who are perhaps too confident that they have seen it all before, that nothing fundamental can change in governance, that in every other generation there is a wave of interest in techniques to support policy making which invariably comes to nothing. This book aims to steer a middle course between the merely cynical view that nothing can or does change in politics, and the romantic view that with only a little effort to master some tools, technical rationality and efficacy in problem solving can be achieved and institutionalised in policy making. E-governance will not, of course, take the politics out of politics, nor should one try to do so: technical rationality and the effectiveness of policies are never the only overriding values in policy making in any society, nor should they be. On the other hand, when the tools that politicians and public servants use in governance change, there are opportunities for doing some things differently – both for better and for worse. New tools can enable people to ask certain kinds of questions that could hardly be raised previously and were only answerable in general terms and with difficulty, but they can also lead to certain other and perhaps pertinent questions not being asked or only being asked too late or too early.

As some leading analysts of political decision making have argued (March and Olsen, 1976; Simon, 1997 [1945]), the real significance of changes in the technology with which decision making is supported, is not that it allows ministers and chief officers to access new and more detailed kinds of information, although that may also be true. For senior policy-making officials do not necessarily want even more detailed information. They face a deluge of information – and, despite the popular discussion of 'information overload', this is not a new experience or an artefact of the coming of electronic mail: it has been the experience of leading politicians from ancient times. Rather, they are constantly searching for ways to filter out less important information, to find in the most useable and concise forms, the minimum of information necessary to make reasonable, robust decisions. Practical judgment must always be an activity that centrally involves information rejection: the key challenge is to reject information more rather than less intelligently, in order in the first instance to distribute the attention of policy makers more appropriately. For, in reasonably prosperous states, the really scarce resource in politics is not the taxpayers' money: it is attention to problems (March and Olsen, 1976; Simon, 1997 [1945]). Attention to problems is allocated in complex ways. It is bartered, bargained, stolen, seized by and between policy makers, forcibly directed by constitutional rules, begged, borrowed or intimidated by pressure groups and special interests, blown hither and thither by shocks, and also, with quiet insistence, steadily attracted in particular ways by the implicit biases in the technologies as they are actually used by policy makers to support them in decision making.

New technologies used in governance throughout history have changed the way in which attention is allocated. The new cartography that supported the decision making of Renaissance princes helped draw their attention to new kinds of governance problem and helped lead them to new kinds of questions about both foreign and domestic policy. The 'avalanche of statistics' that was the greatest of the new technologies of governance in the eighteenth century, drew the attention of Prussian absolutist monarchs and Scottish city élites in new ways to new kinds of problems (Hacking, 1990; Foucault, 1991). In the nineteenth century, the emergence of political economy provided a kind of technology through which new kinds of questions could be, and were asked, about the governance of the means of prosperity (Polanyi, 1944).

To speak in this way of the technologies themselves exerting such powerful leverage over attention is, of course, elliptical. More exactly, the technologies provided the means, the opportunities, the props and the prompts by which policy makers could, given the institutional and political situation they were in, ask new questions and shift their attention. In the same way, the adoption of new digital technologies of e-governance are likely to provide the occasions, the foils, the languages, and the insistent reminders which will enable politicians and the public servants who advise them, for shifting patterns of attention, and consequently for what, in this book, I shall call shifting styles of political judgment.

The central skill that must be cultivated and deployed by the policy-making community as a whole is that of political judgment. Understanding how political judgment works and what role technologies can play in its cultivation and exercise is central to the task for this book. Later chapters will expand on what is meant by this. Yet, even in its simple and rough sense of the astute appreciation of situations and shrewd exercise of choice of how to act or how exactly to remain inactive, it describes the key virtue that policy makers try to cultivate in themselves, and one key yardstick that parties and publics use to appraise the merits of leaders and representatives. The really important question to address to the new technologies of e-governance, then, is this: how can or might they be used to support political judgment, or, more accurately, how might they be used to shape and style it, just as cartography, the production of statistics and political economy were used to stylise political judgment in previous centuries.

I take it that, despite the pressures of institutions, interests, prevailing ideas, biases designed into technologies of e-governance themselves, the ways in which politicians and public officials develop styles of political judgment are not predetermined. As Marx famously put it, people make history, but not in circumstances of their choosing.

The central purpose of this book is to raise in the reader's mind the following question: how can policy makers, constrained as they are by a variety of institutions, today make the use they might wish, should they reflect on it, of the new technologies of e-governance? Or, making use of the available menu of e-governance tools and systems, how far can and should they stylise their political judgment in the first decades of the twenty-first century?

Later chapters of this book will set out both my understanding of the constraints upon the menu, and my conception of the space for choice. The argument will be that the principal constraints that really shape what is done with technologies of judgment are to be found in the *styles of social organisation* that govern the relationships between policy makers.

Some definitions are perhaps in order at this point. When describing and explaining these styles, this book will speak of 'institutions'. By an institution, I mean a formal or an informal constraining social rule, convention or norm (North, 1990, 3), which structures interaction, which is recognised by those subject to it (Knight, 1992, 2), as creating an accountability or some duty to provide an account of performance and be subject to appraisal and either sanction or reward, however informal (Douglas, 1980, 1986), and which leads to the forming of more or less stable social patterns (Jepperson, 1991, 145) (for a review of definitions, see Peters, 1999). Institutions, as the term is used here, are not confined to explicit rule books or constitutions, but encompass informal, tacit and even unacknowledged rules. We can distinguish between empirical institutions – those rules that are explicit and recognised by those subject to them – and fundamental institutions – which are often implicit and informal, and causally important for whole policy or social domains and which define sociometric structures of ties and accountabilities for the basic forms of solidarity that sustain economic, political and social life. Institutions are to be distinguished from organisations and from social organisation: an organisation is a set of empirical institutions bounded by membership, foundation and dissolution and with at least one explicitly prescribed purpose which is supposed ideally to govern the collective action of those members. Institutions in this broad sense are critical to the policy process, for they define the informal accountabilities that structure policy making – for example, they specify with either more or less imperative weight, just which civil servants may share information with which others, which politicians may have access to what information systems run by the executive, how people feel the need to withhold information from others. Institutions are not just rules in the sense of abstract prescriptive ideas, but practices that are more or less entrenched or established in the organisation of spaces and artefacts (such as technologies) used in decision making, and this entrenchment is achieved in ways that give them their material character. By encompassing informal institutions, it can more readily be seen that institutions are necessarily embodied in *material* culture such as the design of buildings and the layout of rooms and the design of physical artefacts including the images that information technologies are configured to present to policy makers, the gait and demeanour of people, the bodily manner of performance. For only through these material forces can institutions communicate and sanction accountabilities, and only through such things can people fulfil or renege on those accountabilities. This recognition will turn out to be of central importance for the argument of the book.

Social organisation is the structure of all the institutions, both fundamental and empirical, that define the accountabilities by which behaviour is ordered. Institutions are not found singly but in sets, which work together (more or less consonantly, if the systems are viable). Finally, it is necessary to speak of *forms or styles* of social organisation, by which is meant the distinct syndromes of consonant fundamental institutions each of which stylises the ordering of behaviour differently. In this book, it will be argued that forms of social organisation among policy makers are the fundamental forces shaping political judgment and the use and meaning of e-governance tools.

For, I shall argue, political judgment is never simply an individual skill, possessed in greater or lesser degree within the brain of the particular politician or public servant (contrary to some psychological approaches used by some analysts of policy making, e.g., Renshon and Larson, 2002). Rather, it is intensely social; it is always an achievement of the style of social relationship among groups of politicians and public officials. By this, I do not mean that it is typically collaborative and collegial: indeed, generally, it is not. Collegial relations are only one type of social organisation, and a rather special type at that. Rather, I mean that the structure of social relationships, however apparently strained or even conflictual, is key to understanding how policy makers develop and exercise judgment, and indeed the organisation of policy makers itself shapes the style of judgment. This has the most immediate consequences for the ways in which policy makers need to think about the software they choose and use: for only some patterns of social organisation among them will sustain the effective use of certain types of software. Second, these styles of social organisation shape the questions that policy makers ask of the technologies at their disposal. Third, this calls for a way of classifying these styles of organisation, in order to help those choosing and using e-governance technologies to select the right software horse for the particular policy course.

Perhaps, however, the course itself is being redesigned? The book will argue that we can best understand changes in the organisation of relationships between politicians and senior officers that sustain their political judgment, by showing how e-governance technologies are deployed by policy makers in their differing organisational situations under particular constraints of, for example, constitutional law and implicit assumptions about how political behaviour is acceptably conducted, to pursue particular projects.

The central prescriptive argument of this book, then, will be that policy makers should begin to think through their use of the new technologies by thinking first about themselves as groups, and as groups of policy makers in a system or network of relationships with other policy makers – ministers, regulators, legislators, local and regional policy makers, party faction leaders, professional technicians of policy analysis in the civil service, and in independent institutes, and perhaps even journalists and other scrutineers.

The ways forward with which this book concludes flow directly from this general line of argument. Used in particular contexts, e-governance technologies can be valuable for policy makers. They can help them to ask fresh questions, to look at problems differently, to draw upon each other's knowledge differently, and to influence the way in which they work together, affecting both the content of their decisions and the depth of their commitment to the choices they make and to implementing them. As other programmes of e-government, in services, democratic life and administrative management generate huge stocks and flows of data, e-governance tools will probably come to seem the most obvious and immediately appropriate ways in which policy makers can make whatever level of use of these banks of information they find most appropriate and helpful. Put another way, gradually, other e-government technologies will come to exert a kind of institutional pressure for the development of e-governance. E-governance tools will be used to support innovations in the management by executive administrations of, for example, budget setting and in the conduct of scrutiny and oversight by Select Committees and by scrutiny committees over the operations of the central and local executives. These innovations may well prove more convenient and tractable for policy makers and perhaps for publics. Scrutiny bodies, the media and the public will increasingly expect policy makers in executive roles to be capable of defending their decisions using the means available to them through e-governance, even though the technologies themselves will never mandate a particular decision.

However, in this book, I am very careful not to oversell these systems. They will neither deliver the policy wonk's New Jerusalem of substituting technical for political rationality, nor should they. Technologies do not deliver once-for-all improvements in rationality, as if rationality were the only thing of importance in policy making, or indeed a single thing at all. What can be hoped for from intelligent use of technologies is the development of disciplines, occasions and prompts for asking more relevant questions and interrogating information more constructively. Contrary to the arguments of some advocates of e-governance (e.g. Cloete, 2001a; Landell-Mills, 2003), there is no realistic prospect that using these technologies alone or even that using them well in the sense of simply increasing the levels of individual skills among policy makers in the manipulation of the software, will reduce the numbers of cases of policy failure or unambiguously improve accountability.

Nor should anyone expect the end of face-to-face communication among policy makers. All digital technologies, and e-governance technologies more than most, are effectively used only in the context of social relationships between the users that continue to be built in face-to-face encounters. It will always be true that in politics, one politician needs, as Israel Putnam put it, to see the whites of the other's eyes, and, as John F Kennedy graphically described it, to see who blinks first.

The challenge to the information technology industry is not simply to develop affordable, easy-to-use products, and to avoid overselling and excessively technical rationalistic assumptions in the design of new e-governance systems, although these are of course important tasks. The central challenge for the industry is to design for the actual forms of social relationships between policy makers, to start where the policy makers are and want to be, rather than to start from where one might ideally want them to be. Policy makers will, no doubt, change the way in which they work. They will do so, however, not in order to make the best use of particular digital technologies, but for fundamentally political reasons. There has been a tendency among some e-governance specialists to explain the slow take-up of many e-governance tools by politicians, in particular (by comparison with some professional civil servants and senior officers), on the grounds that there is something defective about some politicians in particular or about political decision making in general, that means that they do not deserve what they do not seem to want. This is misguided in several ways, not least on commercial grounds. Politics is a highly rational business – in some ways, perhaps too much so, as Thucydides and Machiavelli colourfully pointed out. But its rationalities are plural, and some of them can be quite different from that imagined by those looking at political issues as technical problems of policy engineering and design.

Yet, policy making does shift between these rationalities, and technical change in the means by which judgment is sustained is, as I have argued already, often the occasion, the foil, the opportunity, and the focus for shifts in the style of political judgment. The really important issue which this book raises and attempts to explore, is what kinds of shifts e-governance will be used for.

Finally, e-governance and political judgment are not exclusively the concerns of rival partisan political élites and the most ambitious salaried public servants. In a democratic polity, there is rightly continuing pressure upon those élites to make the style and the content of their political judgment ever more transparent, ever more open to influence from both majorities and minorities among the public, and thereby ever more subject to the accountability of the ballot box. E-democracy and e-governance are, if only every few years, powerfully related to each other. In the periods between elections, that ongoing relationship cannot be confined simply to the processes whereby policy makers gather views as inputs for the deliberative processes supported by the tools of e-governance. Nor yet can it be the role, in a democracy, of the élites to exercise entirely without constraint an authoritative and paedagogical influence over the political judgment of the electors (Kant, 1983 [1784]; Paine, 1988). On the other hand, it is on the one hand, an unrealistically romantic view of democracy, and yet also, too often, a dangerously illiberal one that runs the risk of tyranny of the majority (Mill, 1972 [1861]; De Tocqueville, 1998 [1835 and 1840]), to imagine that popular political judgment can wholly be substituted for that of elected politicians and their salaried public officials as advisors. A decent and politically viable liberal democratic order requires mutual influence between democracy and governance, with the proper right and duty of policy makers to listen, persuade and judge, but with the sovereign accountability remaining with the public (Burke, 1981 [1780]).

These may seem overly abstract and general principles for a discussion about software. Far from it. For they go to the heart of the debate about e-governance and democracy. For to the extent that tools of e-governance enhance the capabilities of policy makers for coming to judgment and for justifying their judgments to their publics, the question becomes the more urgent, how far and in what ways should those capabilities be held to account? The exact definition of just what information from the deliberative processes of governance should be made public and at what point in time after its production is bitterly contested today in the context of freedom of information laws. These conflicts come into sharp relief in the context of the development of digital models, not of policy problems but of the preferences of policy makers as they are in the course of development, and more recently over the rules used to filter messages that politicians design into their electronic mail handling systems, just as they have done for some time in the context of the written and oral policy advice of civil servants to ministers. In the final chapter of this book, these issues will have to be addressed squarely.

Outline of this book

The argument of this book is organised as follows. Chapter 2 introduces the concept of e-governance, contrasts it with the other three main programmes that together make up e-government, and gives a brief history of its development, before presenting a classification of the main types of e-governance tool, according to the function they serve.

Chapter 3 is devoted to more detailed consideration of particular e-governance tools, and examples of initiatives using some of them are presented in a series of boxes containing brief case histories.

Any consideration of the prospects for e-governance must address the question of why take-up of at least the more advanced tools has been so slow, and whether there are reasons to think that there will soon be more extensive adoption. This is the task for Chapter 4, which reviews five types of explanation, showing the fifth to be the most important, before exploring briefly five forces that are likely to increase the demand from policy makers for digital tools to assist them in their work.

There has been a debate for some years about what the consequences might be of increasing use of e-governance tools, and four basic accounts have been offered. Chapter 5 considers these accounts in turn, and argues that the first three, fashionable as they continue to be, have little merit. The chapter argues that the consequences of e-governance will be, in fact, the consequences of the institutional commitments that technologies will be used to serve: the technologies can, in principle, be used for many different purposes.

Chapter 6 presents three case studies in more detail than was possible in Chapter 3, showing how the use of information technologies to support styles of political and policy judgment is shaped by the institutional characteristics of the organisation of policy-making bodies and their relationships with other bodies in the policy process.

The central argument of the book is that we should appraise e-governance tools by the extent to which they are embedded in disciplines and techniques that can contribute to the sustaining of political judgment. Chapter 7 examines what is meant by political and policy judgment, and offers a new account, distinguishing five key elements that make it up.

The first five chapters present the reader with a series of promissory notes, which are redeemed in Chapters 8 and 9. For these chapters provide the account of the variety of institutional settings in which different kinds of policy makers find themselves, and show how these settings shape the styles of political judgment, as measured by the relative weight given to the elements identified in Chapter 7, that each will give. In effect, Chapters 8 and 9 present a new and distinctive account of the policy process, which is designed to enable the reader to see which policy makers are most likely to be interested in which of the types of e-governance tool distinguished in Chapters 2 and 3 and why, and what the consequences of their use of those tools might be.

Chapter 10 is given over to drawing out the implications of the argument of the book as a whole, and specifically of Chapters 8 and 9, for reform strategies to make the best use of e-governance tools. There, recommendations are presented and justified, in respect of the general principles that should govern the introduction of e-governance tools, specific considerations for the legislature, the executive and for local government. Finally, the chapter addresses the challenge sometimes made that providing the policy élites with e-governance tools may have the unwelcome consequences of empowering them to exercise judgment, but at the expense of buttressing their democratic accountability: that section sets out some general principles and some specific recommendations on how programmes for e-governance and e-democracy should be related to each other. Chapter 11 is a brief conclusion drawing together the main strands of the argument of the book as a whole.

2
The Scope of E-governance

In this chapter, the reader is introduced to the main types of e-governance tools now commercially available: the market is so vast and changing so rapidly that I make no attempt to pick out individual branded products. (For a discussion of some branded products, from a political science perspective albeit one that suggests with some qualifications, contrary to the argument of the present book, that the use of these technologies can itself transform the quality and rationality of the policy process; see Cloete, 2001a,b,c.) Rather, the chapter begins by offering a brief definition, and situates e-governance thus defined by contrast with the other programmes that together make up 'electronic government'. After a very rapid canter through the recent history of e-governance since the development of the modern digital computer, the chapter then offers a classification of tools by the principal uses to which they are put.

Defining e-governance

E-government should be divided, in the first instance, into four distinct areas of activity, namely e-democracy, e-service provision, e-management and e-governance; although they are distinct, there are important relationships between them.

By *e-democracy*, I mean not simply the activities of registering to vote and casting a vote across electronic networks, but the *conduct over such networks of the whole panoply of activities by which government solicits, or receives unsolicited, the views of the citizens and indeed of businesses and other organisations on matters ranging from full-scale legislative change to the tweaking of the management of services and programmes to meet the concerns of current consumers.* Certainly, e-democracy includes the management of formal consultation with panels of individuals and organisations, but it also covers the management of satisfaction surveys and complaints.

E-service provision is concerned with the *delivery of public services over digital networks and media* – whether by government bodies or by private bodies under contract or licence – to individual citizens and organisations, including the management of transactions and payments, application for and issuing of permits and licences, the provision of both general information and individually tailored advice, the requisitioning of information from citizens and businesses for tax and regulatory purposes, and the back-office processing, management, mining and warehousing of data captured from these services and transactions in order to develop the service offering.

E-management is that field of activity which is undertaken *using digital means, to handle the allocation of resources across activities within government* within the parameters of what can be considered, at least loosely, as existing policy. That the distinction between policy and management is necessarily blurred, invariably violated in a crisis, and that the reality of the implementation of policy depends upon the exercise of unavoidable discretion by street-level bureaucrats, is certainly not to be denied (Lipsky, 1980; Porter and Hjern, 1981; Yanow, 1996). Nevertheless, in many cases of routine operation – or 'normal science' (in Kuhn's 1970 [1962] terms) – of public management and administration, there is a rough-and-ready distinction between the changing of the goals and the content of programmes on the one hand, and the use by managers and their support staffs of project management tools to manage the allocation of monies and staffs already voted and authorised, the use of spreadsheets and accounts packages for routine budgeting, of e-mail packages and simpler forms of groupware such as Lotus Notes™ for internal communication, the use of presentation packages for the conduct of management meetings, the use of statistical packages for the analysis of stocks and flows of resources and the analysis of performance and audit data. Unavoidably, and not least in crises, the boundaries between e-management and both e-governance and e-service provision become blurred, but – like many distinctions in social science – the distinctions remain useful for many purposes.

Finally, by the term '*e-governance*', I mean to pick out the activities of *digital support for policy formulation and the scrutiny and oversight of the achievement of policy goals*. This includes formal modelling of policy problems, the use of formal meeting management tools, the development of banks of policy-relevant materials in knowledge management systems dedicated to supporting elected ministers and mayors, and indeed encompasses some of the more strategic work commonly conducted under the title of 'performance management'. However, many of the uses of very simple tools such as secure and general public Internet e-mail systems or communications between policy makers are also matters of interest from the point of view of understanding how electronic tools are used in supporting policy work.

These four activity areas cannot, of course, be designed or function effectively wholly in isolation from one another. E-management and e-governance must draw upon data gathered in the course of e-services and e-democracy programmes. Many e-service programmes are best designed to permit citizens, at the same time as they avail themselves of what is on offer, to offer the kind of feedback which is the domain of e-democracy. In this respect, therefore, what is needed is indeed some measure of systems integration. However, systems integration needs to follow the peculiar patterns of inter-organisational relationships that structure the flows of initiative and information between the four functions.

A brief history of e-governance

In the early decades of computing, perhaps surprisingly, e-governance was the area in which most progress was made.

The development and relatively cheap availability of spreadsheet software had an enormous impact on the process by which budgeting was done in government, initially on mainframe systems in central government in the 1970s, and then in the 1980s, on PC-based systems at every level including the smallest local authorities. France was well ahead of the United Kingdom when by the mid-1980s many local authorities were using the 'SIAD Mairie' systems which provided at least finance directors – rather few elected politicians used such systems themselves in those years – with an integrated financial management, project planning and transaction data interrogation system with an underlying expert system engine for modelling alternative budget scenarios which allowed a variety of scenarios to be tested on a wide variety of data and projections, including non-financial data where relationships could be identified or modelled (Klein *et al.*, 1991). At least, these tools should have enabled policy makers to ask the kinds of questions that would make it possible for them, if they had the political courage, to explore, propose and justify larger than normal scale changes. Of course, there were many other factors at work in those years that also put pressure on governments to raise incomes, invest capital and spend differently.

Policy makers were able to use such tools more quickly than previously to compare costs and expenditures, assets and liabilities in a variety of different ways, and to run projections based on different assumptions. Spreadsheet technology at least ought to have provided no more than a means by which those pressures could be responded to, and by which policy makers could gain a better understanding than traditional paper methods offered, of the options that were believed to be available and their implications. It seems reasonable therefore to hypothesise that, insofar as any one development can shake the long-standing and institutionalised tendency of governmental bodies to make budget changes incrementally, this development should have equipped politicians willing to do so to make tax and spending decisions of a more radical kind. This has been argued mainly on theoretical grounds but there are some empirical studies suggesting that less incremental budgeting can be observed during this period (see e.g. van de Donk, 1998). The thesis requires more testing: to date, rather little work of a cross-national nature has been done to evaluate the success, even in their own terms, of the reforms to the budget process adopted across the developed world in the 1980s and 1990s (for US studies, see e.g., Caiden, 1998; True *et al.*, 1999).

But the spreadsheet was by no means the whole story. Techniques of modelling and simulation brought new analytical capabilities to economic policy makers in the post-war years. Indeed, after the military applications, probably the next major category of central government use of computing power in the age of the punched card and the mainframe was the running of assumptions on economic models. Today, the British Treasury's model of the British economy is available on the World Wide Web and analysts can run their own favoured assumptions on it and see what consequences it would project.

From the 1960s onward, analysis has been conducted electronically of data captured from transaction processing systems in, for example, social security, immigration and other fields, in order to alert policy makers to trends, exceptions, anomalies, patterns, which can at least stimulate further questions if not always rigorously test hypotheses. Spreadsheets and statistical packages have been used to construct scenarios and projections from these kinds of administrative and performance data, to support decision making. Early commentators had high hopes that the widespread use by salaried professional policy analysts of such systems would herald an era of more sophisticated, better informed and more rational policy making (Stevens and McGowan, 1985, 177–83).

By the early 1980s, modelling and simulation was being attempted in the field of environmental policy making, as evolutionary change models came to be tractably modelled in the 'artificial life' tradition (Ward, 1999), at the same time as environmental policy makers began to demand systems by which environmental impact assessments of proposed initiatives could be undertaken.

Expert systems were first used in such policy making contexts as social security benefits in the 1970s in the United States and then employed on a much larger scale in the United Kingdom in the second half of the 1980s, in order to test the consistency of current regulations, identify anomalies and vague areas, and to explore the implications for proposals to change entitlements and help policy analysts prepare instructions to those Drafting laws (Portman, 1988, 77–9). Today, these systems are used by every front-line benefit officer to calculate entitlements on individual cases, and by every Citizen's Advice Bureau worker to provide entitlement advice, but their first uses were policy analytical in nature rather than being designed for service provision. Expert systems to model British immigration and nationality law were developed in the 1980s by Professor Robert Kowalski of Imperial College London, using the recursive programming language, Prolog, but these systems were not extensively used inside the Home Office (Frøkjaer, 1989).

Models based on neural nets only came into use in government at the end of the 1980s initially on a modest scale, and then principally in assisting professionals in such fields as public health epidemiology, civil engineering and some technical aspects of financial management, to analyse and diagnose complex systems.

Moving for a moment to a level that is at least in analytical theory below – if not always in practice separable from – that of policy and strategic decision making, by the 1980s, management and case decision making in government was beginning to be able to draw upon more sophisticated tools. For example, by the end of the 1980s, central police agencies were being equipped with quite sophisticated modelling systems for criminal profiling and analysis of data in order to support detective work on particular investigations. Electronic document interchange (EDI) systems were introduced to handle legal, financial and procurement systems, and some systems were designed to support extensive integrated analysis and oversight of flows. Some departments have experimented with shared work space systems and accounted document flow systems to track the movement of copies of documents between individuals, identify editing changes made, alert owners to documents that have not been edited or passed on, and so on. For example, there are studies that suggest successful experimental use of such systems in some German federal ministries (Prinz and Syri, 1997).

Geographic information systems (GIS) have been developed since the mid-1980s to bring together, ideally from a small number of visual cartographic images, hypermedia databases of information relating to territories and localities. Data packaged in such systems include census, epidemiological, economic profile, land use and environmental information. These have been used by policy makers for various purposes, but the first and perhaps still the most extensive use is in local land use planning. These systems have also been used imaginatively for public consultation, by enabling the creation of spatially organised databases that represent public and lay knowledge of environmental processes (Cinderby *et al.*, 2002; Forrester *et al.*, 2002). More sophisticated systems are integrated with modelling tools that allow causal relationships to be explored, and there are some innovative examples of their use in decision making at the policy level in the United States. For example, one review of innovations in governance in the US reports that the police department in the Florida city of St. Petersburg is now using GIS to explore and test explanations of patterns of crime and to model the potential impacts of alternative resource deployment strategies in policing (Borins, 1998, 132–3). The popular game 'Sim City^{TM}' has set a standard and an aspiration for many local politicians for the kinds of GIS that they would like to have, but one that is difficult to meet in practice, because of the complexity of dynamic integration with economic modelling and large data sets that cannot readily be gathered and maintained by local authorities.

In support of communications between policy makers, electronic mail systems were introduced in many departments and authorities in the early 1980s, using a variety of mutually incompatible systems and standards. Only by the beginning of the 1990s were most pre-Internet standards abandoned. Secure but simple e-mail listserv systems began to proliferate to support idea sharing among policy makers. Videoconferencing systems were in modest levels of use in some British central departments such as employment by the end of the 1980s.

Intranets came into general government use during the first few years of the 1990s and while take-up was high in some fields of government, actual use was typically of a limited subset of the features of such systems. In many departments and authorities, intranets were often used as little more than over-engineered internal e-mail systems (National Audit Office (NAO), 1999). Off-the-peg groupware such as Lotus Notes^{TM} is used in some parts of government – and is a major commitment in reinvention programmes in states such as Victoria, Australia (see <http://www.mmv.vic.gov.au/>) – but these systems are too often only used for very simple shared data requirements such as external contact management.

The development of the World Wide Web since the 1980s and its availability to policy staff and politicians since about 1990 has supported those policy makers with the time and inclination to search with varying degrees of difficulty for public information of varying quality with which to help build arguments. By the end of the 1990s, in at least some fields, international agencies offered some accredited and high quality and restricted access materials to policy makers over extranet sites accessible through the web. Some of these bodies have experimented with online electronic conferences run over restricted access websites.

Political parties have invested extensively in decision support systems geared exclusively to political communication, such as the Labour Party's 'Excalibur' system, much vaunted during the 1997 general election campaign. More recently, such systems have been developed within government. The British Cabinet Office has announced that ministers will have access to a new system that will provide best arguments for and salient facts and statistics illustrating all current policies, programmes and performance.

However, backbench MPs in many countries appear from the recently available data to be using few e-governance tools, although many are using customer relations management (CRM) software for handling constituency matters, and some are using their own websites to promote online debate in their constituencies. For example, a recent survey of Canadian federal elected representatives found almost two-thirds of them making extensive use of government web sites and regarding this as critical to their work: the survey also found that typically, MPs received policy information from civil servants either online or in writing rather than using the technology to present their own policy arguments to executive officials (Malloy, 2003).

The menu of e-governance systems

One way to classify e-governance systems is roughly according to the main use or function for which the tools are used; of course, there are overlaps because some tools are put to more than one use. There are e-governance tools for six basic categories of purpose, and in the following listing, examples of the main kinds of tool are given:

Box 2-a   A classification of the types of e-governance tools by function

1. *generating understandings*:
   - simple data systems enabling dictionaries of key terms in the dialects of different policy makers from different professional or organisational cultural backgrounds who are collaborating to understand one another's vocabulary;
   - idea generation tools;
   - graphical problem structuring tools (modelling in software procedures such as soft system methodology, robustness analysis, strategic options development and analysis (Rosenhead and Mingers, 2001 [1989]);
   - mental mapping and mental representation tools that enable users to develop graphical representations of their own or others' basic conceptual approach to problems;
   - online consultations, using, for example, the World Wide Web, to put out documents for unstructured response, or providing structured response formats, or offering electronic conferencing facilities with simple 'threading' structure for contributions to be organised freely as responses to topic-originating contributions;
   - argumentation support tools to help multiple groups of decision makers working simultaneously to generate options, identify pros and cons, track multiple flows of argument and debate (Dennis *et al.*, 1997; Buckingham Shum, 1998; Conklin, 1999);
   - electronic whiteboards for graphical representation of connections between ideas (Massetti, 1998); and scenario building tools;
2. *collecting data* or observations:
   - search agents;
   - digital agents based on neural nets for context sensitive searching or editing or précis or transactions (6, 1999c);
   - sensors; and communication recording and storage systems;
3. *organising and analysing data* on events, conditions, problems, processes that have been observed:
   - spreadsheets and budget systems, one of the earliest areas of large-scale e-governance (e.g. the French 'SIAD Mairie' system of the 1980s: Klein *et al.*, 1991);
   - organisational memory capture and management tools (Buckingham Shum, 1998);
   - document profiling systems in shared work spaces that enable users on saving documents to provide hypermedia linkages to related documents, to identify key relationships with key organisation documents; some electronic document management systems use bar codes on paper documents to enable their linkage with electronic versions (Prinz and Syri, 1997);
   - hypermedia geographic information systems (GIS);

- training simulation systems for decision makers, for example for crisis management; and
- formal models;
4. *supporting communication* and transaction:
   - e-mail;
   - electronic conferencing;
   - video-conferencing systems (Mosier and Tammaro, 1994);
   - meeting management tools (Niederman *et al.*, 1996);
   - tools to model and manage conflict, come to consensus or operate a decision procedure such as weighted voting or arbitration (Watson *et al.*, 1994);
   - argumentation support systems; and
   - electronic document interchange;
5. *modelling decisions and advising on possible consequences*:
   - spreadsheets;
   - expert systems for example, to test consistency and precision in draft legislation as in social security (Portman, 1988, 77–9) or immigration (Frøkjaer, 1989);
   - neural nets (Whitby, 1996; Berry *et al.*, 1998);
   - modelling systems for problems such as criminal activity, integrated with resource allocation tools (Borins, 1998, 132–3); and, finally, there are
6. *environments that provide integration and storage* for the other categories:
   - intranets; however, in many cases only the simplest functionality of intranets is actually used (National Audit Office (NAO), 1999); and
   - the World Wide Web.

Categories 1, 3, 4 and 5 include the tools often called '(group) decision support systems' (GDSS) (Jones, 1994; Karagiannis *et al.*, 1994; Finlay and Forghani, 1998). This category is not wholly distinct from that of 'knowledge management' (KM) tools, but the general difference is that knowledge management tools are supposed to manage stocks of knowledge, while group decision support system tools manage flows.

This broad clustering of technologies obviously allows for some particular tools to be classified in more than one grouping, if they are readily used for more than one of these purposes. However, this classification will be used in this book, because the activities of generating understandings, gathering data, organising and analysing it, engaging in communication among those involved in making policy, modelling decisions and defining working environments are all central to what it is to exercise policy and political judgment. Chapter 3 provides examples of the uses of some of the key technologies in each category.

3
Examples and Initiatives

Using each of the types of e-governance technologies listed in Box 2-a, this chapter presents and discusses a number of cases and initiatives, in which policy makers have experimented with e-governance tools, and some preliminary lessons are drawn. These lessons will be used and integrated into a larger framework for thinking about the institutional conditions for successful use of e-governance tools in Chapters 8 and 9. However, the chapter concludes by noting some important differences between those tools that have been widely taken up, and those that have not. This provides the basis for the following chapter, when explanations will be considered. For it is crucial for understanding the prospects for e-governance that we understand why some tools are used and why others are not, and what factors shape the ways in which policy makers demand and use these tools.

It will become clear that each effective use – indeed, in some cases, any meaningful use at all – of these technologies in supporting policy judgment requires the mastery by individual policy makers or by groups of policy makers of particular techniques or disciplines. These disciplines are not merely intellectual, but often involve quite specific uses of the physical space of meeting rooms and banks of terminals, quite specific structuring of time during the exercise of the discipline, and require certain kinds of relationship between policy makers to be in place before the techniques can be effectively exercised at all. Some of these techniques are fairly straightforward, and can be argued to represent only incremental extensions of the kinds of skills that competent policy makers should be expected to possess in any case, or else – in the case of elected politicians who are not recruited on the basis of prior possession of professional skills in policy work – to be capable of acquiring within a short period after election or appointment to ministerial office. Others are more specific, and require both learning and provisional acceptance of disciplines that will not seem intuitive to a policy maker with a conventional training in such skills as conducting written policy analysis and research, engaging in face-to-face debate, oratory, face-to-face negotiation and conflict resolution, organisational mobilisation, management and leadership. This embedding of technology in technique and discipline is crucial to the understanding of the role of e-governance that will be developed in later chapters.

To the extent that e-governance represents a qualitative change in the work of making policy, it is often more due to the learning and exercise of these disciplines by different groups of policy makers than it is due to the use or availability of the physical hardware or the software. We shall see in later chapters that some of the most important qualitative changes in the work of making policy are found to be rooted in the ritual context in which these disciplines are carried on.

Generating understandings

The first and, as I shall argue in Chapter 7, one of the most important elements in policy making is developing an appreciation (Vickers, 1995 [1965]) of the most appropriate way in which a particular group of policy makers can frame a problem. This is not necessarily or even typically the first 'stage' in any chronicle of a policy issue, for appreciation is conducted throughout the whole policy process. However, it is of overarching importance.

A 'frame', in this context, is both more and less than a characterisation or a description (Goffman, 1974; Goodman, 1978; Tversky and Kahneman, 1981; Gamson, 1992; Olson and Torrance, 1996; Kahneman and Tversky, 2000; 6, 2003). It presents both a basic set of parameters that specify how those policy makers' own interests, concerns, fears and hopes will relate to that problem, it also specifies the bias with which they will – indeed, *need* to – perceive the problem. The framing that is appropriate for a particular group of policy makers will depend on their own history, their partisan or their departmental or organisational allegiances and on their situation. Any policy issue can be given many frames, each of which is equally appropriate for a particular group. For example, the challenges of mobile telecommunications regulation can quite appropriately be (in the British context) a technology policy problem for the Office of Science and Technology within the Department of Trade and Industry, a problem of competition policy for the Office of Fair Trading which reports to the same department, a problem of social exclusion as far as the Social Exclusion Unit in the Cabinet Office is concerned, an issue of safety and perceived risk as far as the environmental directorates of the Department for the Environment, Food and Rural Affairs and the public health directorates of the Department of Health are concerned, with their brief to cover questions about the impacts from masts, and so on. Just as, depending on what one needs to do with it, a hammer can be a tool, an archaeological artefact, a piece of evidence in a criminal trial, an example of contemporary design, or a sample with which to test the quality and purity of metal used in the tools industry, so a policy problem can be many things to different groups of policy makers depending on their interests.[2] As we shall see in Chapters 8 and 9, the situation of the policy maker also makes a difference to the frame that is appropriate to them. An issue may be an opportunity to display leadership, a last ditch in which to make one's stand, a stick with which to beat an overweening public authority, a chance to engage in a programme intended to change the cultures of some sections of the public, depending on where one sits, and often this will be at least as important in shaping the frame for the problem as the selection of policy interest. Without such framing, there is no clear question or problem for the policy maker to address, and therefore no clarity about what would count as a more or a less intelligent thing to do. One use of tools, then, is to help people develop ideas about alternative ways of framing a problem, which, in practice, means supporting the use of human imagination, discussion, argument and either agreement or agreeing to disagree and work separately.

Electronic brainstorm support tools, electronic whiteboards, idea generation tools

Three important categories of software that can be used to support the development of appropriate frames with which differently situated groups of policy makers can think about the policy problems they face are electronic brainstorm tools, electronic whiteboards and idea generation tools.

Essentially, these tools are digital extensions of the well-known manual equivalents of the flipchart and set of coloured marker pens, the pack of sticky 'Post-it™' notes, the whiteboard on the wall, or the thesaurus to support free association. The differences are significant, but not fundamental. These systems provide facilities for easy storage and retrieval which their manual equivalents do not; the size of the field is almost unlimited, unlike the bounded space of the A1 flipchart sheet or the 6×3 metre whiteboard; they provide easy ways to mark particular contributions as belonging to particular authors. Electronic brainstorm tools are often used most effectively when they are embedded in meeting management software or at least in asynchronous groupware, because that provides a discipline, a context and a structure and therefore a clear focus that is lacking in the stand alone software for individual use.

Online consultation

The commonest practice of e-governance is, as one might expect, technologically the one that is familiar to most people, the one that represents the most incremental extension of traditional practices using printed paper, and the one that requires policy makers to make the fewest commitments to the public, to each other, or to the future. This is the practice of conducting consultations online, and typically using the World Wide Web.

The narrowest forms of online consultation are those within the executive itself. The Finnish cabinet, for example, uses online consultation among ministers for a great many routine decisions (Paula Tiihonnen, personal communication). Draft or proposed decisions are placed on the ministerial intranet by the minister responsible for initiation; face-to-face discussion at Cabinet meetings is only required if there is an objection from another minister. Since ministers have incentives to avoid putting out proposals that will likely elicit such objections, they try to reach agreement with colleagues in other departments well before this stage. However, all cabinet ministers are expected to keep to the discipline of checking documentation on the intranet, in person, using their own PC: failure to do so results in the loss of a key final opportunity to articulate their departmental interests.

National governments and most local authorities now put huge masses of documentation on their websites, where those interested in commenting can access it, often more cheaply or at any rate more readily than they could buy paper copies. At the most basic level of online consultation, a simple e-mail address is provided for response in whatever structure the respondent chooses to write in. More structure can be created by providing a response format. For example, a form might be provided on the site which requires multiple choice answers, or at least provides distinct sections in which the respondent is asked to enter text in response to particular questions. More deliberative processes can also be supported in association with consultation by hosting an electronic conference, and here the online consultation becomes more like an argumentation support system, albeit with unlimited public access and in public. Typically, electronic conferences strike off from some key propositions or questions derived from a central document on which the authority is seeking to consult, and these propositions or questions are used as the initial contributions to which responses, and responses to responses, are attached in 'threads'; however, in most electronic conferences, participants are given the right to start new threads. In essence, this represents only an incremental extension of traditional paper publishing and inviting responses by letter post and perhaps of providing a telephone line with an answering machine to record brief comments. It is, by comparison with other e-governance options, relatively cheap, and therefore valuable, because governments must consult on so very many initiatives in order to explore with those most affected or interested, how their proposals might be received.

The British government has created a space within the 'www.ukon line.gov.uk' site for citizen access to formal consultations on specific policy proposals and for general comments. However, such open systems do not typically elicit representative samples of opinion, or expertise, and often require moderation of a kind that may sometimes be regarded as unacceptable by some of those whose more offensive remarks are effectively removed from what can be viewed by others visiting the site.

Box 3-a Finland's Committee for the Future

The Finnish Parliament's Committee for the Future is reportedly unique, for it is a parliamentary committee with no direct responsibility for scrutiny of any branch of the executive, no direct budgetary oversight, but has a purely deliberative role to examine long-range challenges for the national agenda and to produce analytical reports exploring the implications of scenarios and projections for the medium term future. The first committee was created in 1994 and the second was established in 1996 and lasted until the 1999 elections: the third committee became a permanent body after those elections. It has conducted enquiries on the implications of forecasts of a wide range of technologies, studies on lessons to be learned from Singapore and South Korea (Committee for the Future, 1996), China (Committee for the Future, 1997a) and Japan (Committee for the Future, 1999), hosted conferences including one of the implications of the internet for politics (Tiihonnen, 2000), and several studies on the future of work (Committee for the Future, 1997b, 1998, 2000a).

The Committee has made extensive use throughout its work on the technology of videoconferencing, often setting up the videoconferencing equipment in the main auditorium in which the Parliament sits. Each of its studies on lessons from east Asian countries and from Wisconsin in the USA were conducted using videoconferences and video seminars with speakers in the states under study. The formats used involved both formal presentations, question and answer sessions and round table deliberations. A less formal structure was used in a videoconference with the Finnish software engineer now based in California, Linus Torvalds, who developed the globally successful operating system, Linux (Committee for the Future, 2000b). In that case, experts from around the world were invited to take part in the videoconference discussion from their own workplaces. The most ambitious use of videoconferencing by the Committee was the experiment in conducting a videoconference over the World Wide Web on an extranet site with 2000 legislators, embassy staff, invited experts, think tank staffers and others from around the globe in 1998. The conference lasted eight hours, and resulted in a series of specific recommendations from the Committee to the Finnish government.

The decision to make heavy use of videoconferencing reflects a view among committee members of the importance of less emphasis and reliance upon document-based forms of deliberation, and the importance of verbal communication. Dr Paula Tiihonnen, the Committee's principal researcher and special advisor and author of many of its reports, points out that as international communication and learning become more important for policy makers, the importance of verbal communication increases because of the slowness and difficulty of textual translation. Moreover, she argues, in verbal communication even with simultaneous interpretation, it is easier to build trust and the kinds of personal relationships, however causal and fragile, upon which successful information exchange between policy makers so often depends (Tiihonnen, 1997). That politicians would recognise this, given the importance to them of rhetoric and oratory, and of connecting the activity of storytelling in leadership to their own faces and identities, is perhaps not surprising. Nevertheless, it certainly represents a departure from the norm for a committee of a legislature.

A commitment to greater public participation has been central to the manner in which the Committee conducts its enquiries. One of the initiatives of the first Committee in 1994 was a televised discussion, again in the Parliament's auditorium, in which Finnish citizens could and did take part (Arter, 2000, 5). More recently, the Committee has experimented with the involvement of its members in web chat rooms for short periods, around the time of municipal elections. Two citizens' juries have been recruited, one comprising exclusively young people and one of older citizens, to work almost exclusively online to provide deliberation and commentary on the reports and recommendations. The Committee is now conducting a series of online discussions using their website specifically aimed at young people (Arter, 2000, 14). This hardly seems unusual in a country in which mobile 'phone voting in mass opinion polling and advisory referenda organised by an independent public commission is common place, popular and influential with policy makers (Tiihonnen, 2000).

The website for the Committee can be found at http://www.eduskunta.fi.

Source: Author's interview with Dr Paula Tiihonnen, November 2000

Online consultation is, however, more often conducted by policy makers with narrowly defined groups of individuals, experts, other policy makers, foreign experts and so on. For this purpose, videoconferencing is often a highly suitable tool. Perhaps the best known recent example of this type of electronic consultation is that conducted by the Committee for the Future of the Finnish Parliament (or Eduskunta).

Argumentation support systems

Electronic argumentation systems use groupware to support deliberation, the gathering of ideas, suggestions, debates, disagreements. Typically, they use software that allows the 'threading' of particular strands of argumentation in ways not dissimilar to the way in which Usenet works, although typically, they provide additional functionality. For example, they allow anonymity to be used (and turned off); graphical representation of threads is supported; certain types of summary statistics can readily be run on the state of debate, and so on. Argumentation support systems can be used either in a dedicated meeting room or remotely, so that policy makers can take part – usually over longer periods than when in a dedicated room together – from their homes, hotel rooms, airport lounges, and so on.

The use of computer supported collaborative argumentation (CSCA) is still not widespread even in the private sector (Bannon, 1997; Conklin, 1999). In public policy making, more use of it has been made in the rather limited implementation of online electronic conferences run over several days, in which participation is at a distance, conducted over websites, and where participation tends to be voluntary, and where no critical decisions hang on the work done. Many such exercises have been run which have involved British politicians and senior civil servants. Little that is of general import can be learned from these initiatives, except that a limited subset of the disciplines of using argumentation support may be diffusing among some policy makers. This is important, because many of the best-known tools in this field have suffered underuse because of the training demands upon users, as much in the logic they use as in the manipulation of the software itself (Conklin, 1999).

However, the asynchronous character of contributions of such programmes can have its uses, where it is not possible or desirable to work simultaneously. However, it is hard to extrapolate from these online conferences as to what might happen if policy makers were to make more disciplined use of such systems in settings where key decisions are to be made, in which participation is compulsory, and where substantive disagreements actually matter.

There are also important synchronous forms of argumentation support. These systems use groupware to manage meetings, often within the parameters of a particular discipline or technique. Box 3-b describes a revealing recent initiative in the United Kingdom.

Box 3-b Decision support conferencing in a British all-party group

One very well-known British business school professor worked with a prestigious all-party group in Parliament to conduct a decision support conference. The all-party group had been conducting a Foresight exercise, looking at a range of organisational, technological, policy and regulatory trends. This had involved a consultation with experts in business, government, pressure groups and academic centres. The consultation had thrown up more than two hundred propositions, many of which were inconsistent with each other, and each of which raised several policy challenges. The all-party group faced a major challenge in whittling down the number to something manageable, as well as securing agreement among their members about the most likely trends and implications, the interesting ones, and therefore in producing a report that would have the necessary focus to be influential with the executive.

The decision support conference was held in Whitehall in just a single day. Fifteen members of the all-party group took part. The business school provided its portable system for decision support conferencing and a facilitator. The hardware consists in a set of personal computers, a network and the software to support structured argumentation and to allow simple polling and analysis of expressed preferences to measure the extent of consensus within the group. The room was laid out in cabaret style, to encourage teamworking, and to permit a greater sense of participation than is possible in theatre or boardroom style. As we shall see throughout this book, the design and the use of space to manage the process of judgment is a crucial part of most disciplines in which technologies can most effectively be used.

The conferencing software is a group electronic meeting support system. (For discussions of the uses, aspirations, limitations and lessons from experience in using such systems, see e.g. Weatherall and Nunamaker, 1995; Nunamaker *et al.*, 1996.) It provides support for participants to engage in argument anonymously, to offer ideas, and to elicit measures of agreement and disagreement, support and rejection, interest and lack of interest, in the propositions put forward initially or else that emerge in the course of the conference. Most importantly, this style of decision support conferencing enables all the participants to contribute at the same time, without either waiting for others to finish their contribution or being so rude as to interrupt them: in short, it supports parallel working. This can be used in ways which save time, and where time available is short, as it was in this case, this can be very valuable.

The aim, for which the facilitator is responsible, is not to use the voting as a definitive ballot but to use it to measure the spread of views, to identify extremes, and thus enable priorities to be set for the use of debate and discussion time.

Systems of this kind work well where there is willingness among the entire group to collaborate together, where all members are prepared to accept the limitation upon their individual ability to 'railroad' a decision through, and where all are prepared to submit to the discipline of the style of management of the meeting that is designed into the system. In this case, success was only partial. Indeed, the task was completed, and the number of propositions was whittled down to less than twenty, which was manageable. However, one member of the all-party group said very little, clearly resented the process, and only spoke to try to use conventional forms of veto that might be acceptable in a standard face-to-face meeting. This member of the group felt no commitment to the process and no 'ownership' of the product, and felt much more comfortable in face-to-face meetings where skills of influence that person possessed in a high degree could be used to their greatest effect. The facilitator was astute to the way in which this person was behaving, and managed to limit the damage that might otherwise have been done to the management of the meeting process. Nevertheless, this did, the business school appraisal team felt, represent a significant qualification on the success of the experiment. They were led to wonder whether such problems are more likely in the highly politicised environment of a group of Parliamentarians or public sector policy makers generally. In Chapters 8 and 9, I shall offer an explanation of the particular situations in public sector judgment contexts from which these problems arise, although I shall not comment on the question of whether in general, such situations may occur less frequently in business settings.

This experiment reinforces the findings of others who have worked with electronic meeting support and electronic argumentation support systems, that, although with suitable participants, there are significant gains to be made in the quality of deliberation and judgment, the discipline involved can represent a major challenge to the style of leadership, management, influence of some participants, and limited success can be expected with such systems if, for any variety of reasons, they are not prepared to adapt to that discipline (Nunamaker *et al.*, 1996). In less time-constrained contexts than this one, other commentators find it useful to run separate track programmes for such persons (Nunamaker *et al.*, 1996, 16), but of course in come contexts, this can frustrate the purpose of the exercise.

Finally, the anonymity of such systems has both advantages and disadvantages (Weick and Meader, 1993). The advantages are that it encourages candour, and many problems are more easily solved when people know what others are really thinking. Unfortunately, the happy Enlightenment view that more mutual understanding is always better than less, is not always borne out in policy making. Sometimes, the fact that hidden assumptions remain hidden is essential to the functioning of a group of policy makers, and one risk in the encouragement of candour is that it can lead to explicitness of the kind that ends collaboration, if the group is not compelled by statutory duty or by political necessity to work together.

Collecting data

Sensors

A sensor is essentially a device that is designed to be sensitive to some kind of event, to capture data about that event in some recorded and storable form, and to transmit it to some other system, where it may be used, analysed, classified, aggregated, inferred from, and so on, or, of course, discarded. In analogue systems, closed-circuit television cameras and cassette recorders with microphones are sensors, as are a great many medical and scientific instruments for detecting, for example, pulse rates or radiation levels. In the electronic context of modern computing, sensors provide information in digital form, which can be stored and analysed more easily. Those aspects of policy making that are very hungry for data can use systems that deploy sensors to good effect, once decisions have been made about what data is required, how it will be used, where the sensors should be deployed, and so on.

For example, environmental policy makers can acquire data about pollution levels in an area by installing sensor devices that are designed to be sensitive to levels of soil nitrate, levels of methane emission or levels of radon. Data captured by these sensors is remitted back to monitoring centres for aggregation and analysis. These analyses can then be used in decision making about the use of regulatory powers in pollution control policy. The quality of those decisions depends not only upon the performance of the sensor devices, of course, but also on the prior decisions about their settings, the target variables and how good they are (and continued to be over time) as proxies for the underlying concerns, and of the decisions about where to locate them. This kind of environmental programme is expensive to run, not least because sensors are expensive per unit, and because they are easily damaged and soon wear out. However, the state of Florida in the southern USA is making growing use of sensors for this purpose, because of the chronic environmental sensitivity of the everglades.

Many public authorities with social service responsibilities are making use of sensors in their programmes of domiciliary care, to monitor the health of, for example, older people, and to alert care workers in the event of a fall. In the criminal justice context, the high profile programmes of electronic tagging of offenders released on bail, parole or licence into the community are also essentially using sensors.

Most uses of data from sensors tend to be applied by operations managers in the course of service provision or micro-level regulatory decision making, rather than by policy makers for larger scale policy issues. However, there are occasions when it makes sense for time series analysis and comparison of data captured from sensors to be used in policy analysis, and it would be reasonable to expect this to be a growing area of use and interest.

Digital agents

Digital agents are pieces of software that are built around neural nets, which are 'trained' by trial and error and experience to perform a wide range of functions remotely over networks. Already, agents are in very widespread use for all sorts of ordinary commercial purposes. One recent review in the trade press offers the following list of purposes (Hunter, 1999):

- gathering and summarising network information;
- controlling software installation, upgrades and bug fixes;
- coping with information overload by filtering out unwanted data and identifying important elements;
- detecting network traffic patterns and anticipating bottlenecks (e.g. in telephone network systems) (Ward, 1999, 223ff);
- scanning websites for specified types of information;
- monitoring specific information sources for changes the user is interested in, such as particular stock prices;
- monitoring software licence agreements;
- delivering new features to programmable devices and even electronic appliances such as washing machines;
- monitoring service-level agreements in both public and private networks;
- tracking progress of IT projects;
- monitoring end-to-end application performance – 'the user experience'; and
- monitoring systems for consumption of resources, such as toner in printers.

It is probable that in the next few years, a wide range of personal agents will be on sale that will provide individuals with the means of putting together a personalised daily news service, arranging personal finances, anticipating desires for purchasing, making transactions, organising one's diary, communicating with other agents to broker arrangements with other people, and so on (Barrett, 1998).

Perhaps the first really sophisticated digital agents to be developed were software viruses. For many viruses are software agents that are smart enough to know how to reproduce themselves, search out particular kinds of information and destroy them, bring them back somewhere, and so on. Indeed, often the only thing that distinguishes a virus from an agent is that a virus is malign and an agent is supposed to be benign, both from the perspective of the principal and of those with whose systems the agent makes contact. In order to protect their integrity against viral attack, many data systems, websites and private networks have been fitted with systems that lock out all agents, good or bad. However, these defences may have to be redesigned as many agents are introduced that are accepted as genuinely useful, and as systems of accreditation, trust and authentication are developed to enable benign agents to identify themselves more readily, but which are more difficult for malign viruses to forge or replicate (6, 1999c, 2001a,b).

There are many potential uses for digital agents in policy making, but their use is in its infancy. The simplest use will be in collecting information from across networks such as the World Wide Web, but probably the most important and also the riskiest use will be in 'training' their neural nets to carry out précis and analysis of materials gathered, upon which policy analysts and policy makers may come to rely.

Organising and analysing data

Geographical information systems

Geographical information systems (GIS) comprise a wide family of software tools by which data can be represented by showing cartographic distributions, divisions, and relationships, and from which a variety of other data, applications and other resources can be attached to particular geographically represented units through hypermedia links. These systems enable the generation of fresh cartographic representations of data within moments, in order to demonstrate the effects of incremental changes in data.

Their uses are many, but the earliest uses in the public sector have been in land use planning and development control, and in the building of more sophisticated registers of property ownership. Increasingly, they are being used to support numerous kinds of decision making. A common indirect use is to support both elected representatives and managers with defined geographical jurisdictions by providing them with ready access to information about their areas. More directly, GIS tools are being used to support decision making in performance management, by enabling cartographic representations of comparison between performance in different areas, by allowing the modelling of flows of resources between areas and enabling colourful and easy to understand comparisons on a geographical basis between needs indices and activity measures for service provision. Of course, these are all forms of analysis that have been used for centuries, but the speed and the flexibility of the map generation capability enables scenarios to be generated and compared much more quickly than by manual methods, and the immediacy of the software-generated representation makes GIS a powerful tool for advocacy.

Two of the case studies presented in Chapter 6 provide more detail on the uses of GIS in decision making in two US authorities. Increasingly, however, GIS systems are being used to support a wide range of decision making in which it is necessary to decide on allocations of resources between different geographical jurisdictions, or to design flows of resources across multiple places (such as routeing of waste collections, and other 'travelling salesman' problems that require mathematical solutions to geographical problems).

Knowledge management

'Knowledge management' is the managerial buzz-word of the decade. It is closely linked with the prevailing commitments in contemporary thought about organisations and management, including the aspiration for 'learning organisations', for capturing, valuing, rendering explicit, retaining and exploiting the ideas and knowledge in the heads of staff, for valuing organisations more accurately by taking explicit account of their intangible assets, for being much more explicit and methodical about the ways in which risks are managed using everything that is known about risks at every level in an organisation, and so on. Within the overall heading of knowledge management, there lie a number of quite distinct activities and functions. These include

a. developing inventories of what is considered to be knowledge in an organisation;

b. identifying individual roles and responsibilities for the production, cultivation, recording and dissemination of knowledge;
c. rendering implicit and tacit knowledge explicit, where that is considered to be possible and appropriate;
d. giving formal structure and organisation to the body of explicit knowledge inventoried and stored; and
e. creating directories of individuals by category of expertise or non-explicit knowledge, which can be accessed internally within the organisation, and, where appropriate, by outsiders.

'Knowledge' here can mean one or more of several quite distinct things, including at the very least

i. skills, or know-how; this knowledge is often tacit or implicit (Polanyi, 1958);
ii. bodies of explicit propositions of a factual or an interpretive character, organised into distinct documents, which may or may not be new, exclusive or proprietary to the organisation, distinctive, or other such;
iii. discoveries, or new, distinctive, exclusive and proprietary bodies of propositions of a factual or interpretive character, which form assets that can be exploited to further the purposes of the organisation.

In addition, although I shall not discuss this in depth here, some uses of the term 'knowledge management' focus on the concepts that people use, and therefore place great stress on 'mental mapping', 'concept mapping' or 'cognitive modelling' techniques, as formal techniques of representing knowledge in 'knowledge engineering'. Although these have great importance in particular studies, their usefulness in the kind of work that policy makers do is not yet established, except as intermediate products or tools or background instruments to help designers develop structures that seem intuitive and recognisable, for hypermedia organisation of bodies of propositional or documentary knowledge.

Bringing together these activities and kinds of knowledge yields a minimal set of possible knowledge management programmes all of which can be used in different agencies within the public sector, in support of at least some aspects of the work of policy-making staff. Figure 3.1 gives some examples.

The simplest and the most common knowledge management activities in the public sector are those shown in cells (i.a.) and (i.e.). Although in some sense, most of the activities in column (ii) have been undertaken ever since archives and libraries began to be kept in government, the

Knowledge type Activity	i. Skills, know-how	ii. Formal bodies of propositions	iii. Assets
a. Inventory	'Yellow pages' of experts by category of expertise	Multi-dimensional classification of all documents created, using document flow and management system	Classification and bank of all proprietary intellectual property generated, held, purchased, rented
b. Roles	Creating a Chief Knowledge Officer, departmental Knowledge Leaders	Creating a policy document management structure, updating responsibilities and division of roles	Creating intellectual property oversight team and lead officer
c. Rendering explicit	Disciplines on all staff to write entries into organisational manuals for their own key skills	Discipline upon key staff of writing down, classifying and storing records of all policy decisions	Standard codification practices for all intellectual property generated or used
d. Formal structuring	Hierarchical order by category, represented in hypertext	Hierarchical order by category, represented in hypertext, or accessible using key-word search engine	Hierarchical order by category, represented in hypertext, or accessible using key-word search engine
e. Making accessible	Publishing a 'Yellow pages' on intranet or website	Document bank on intranet or website	Consistent pricing policy for intellectual assets, reflecting systematic valuation, and publication and sales programme

Figure 3.1   Knowledge management programmes: some examples

digitisation of this process does represent significant work in reclassification and some distinct challenges, as well as opportunities for more flexible and easier-to-use coding and access. Cell (i.c) represents a major challenge in any organisation. In general, experience suggests that only the most disciplined, formal and hierarchical organisations can achieve the rendering explicit of significant quantities of tacit knowledge with great consistency and commitment. For example, in military organisations, there are both carrots and sticks, blunt authority and promotion incentives upon staff to comply with duties to render their knowledge explicit, and to feel that they have reason to be willing to share that information with others rather than hoard it for their own use. Other uniformed services can also draw upon these types of resource and capability, but it is more difficult for organisations with looser, more egalitarian or more individualistic cultures to achieve this discipline.

As we shall see in Chapters 8 and 9, the degree to which organisation members at each level in an organisation have reason to be willing to share information is often the single most important factor in the design of any e-governance programme. Finally, such programmes only work if senior managers and politicians both actually value the tacit knowledge that lies in the heads of their staff at every level, including front line staff, and if those staff believe and feel that their tacit knowledge is valued. Despite the popular managerial rhetoric that 'our staff are our greatest asset', it certainly cannot be taken for granted that this is universally practised in government, or indeed even within policy-making centres in central government departments, in agencies or in local authorities. (This is a limitation in knowledge management in the public sector upon which, in the British context, Milner 2000 lays great emphasis.) It is for these reasons that, outside the uniformed services with their highly disciplined culture, systematic programmes of explicit recording of otherwise tacit knowledge are typically difficult to sustain.

The management of intellectual property in policy-making work (column iii) has typically been a matter of pricing policy for publications of reports, policy papers and so on. There are few cases of policy work in government that are very closely analogous to the management of intellectual property generated in, for example, a biotechnology research company or a software house, where large streams of revenues hang upon major programmes of investment, research and development in proprietary bodies of formal knowledge in new discoveries. (There are some analogies, again, in the military research and development and intelligence communities, in cartography and also in public sector meteorology, but these are not our concern here.)

Box 3-c presents an example of an initiative in knowledge management in government for, among others, policy makers, which has made a start with the most straightforward activities such as those in cells (i.a) and (i.e), (ii.a) (ii.d) and (ii.e), but which has not yet set itself higher ambitions such as those for row (c).

Supporting communication

Groupware

Probably the most widely used groupware tool is Lotus Notes™. The package provides support for a variety of group activities including e-mail

Box 3-c   The Knowledge Network in British central government

The Knowledge Network is a document management system that runs on the British central government intranet system, the Government Secure Intranet (GSI).

Its origins lie in some work done in the Department of Health, to produce an online briefing management system for press officers and policy staff, to support the better distribution of briefing materials. The civil servant responsible for this was initially successful in interesting the Prime Minister's office in a similar system. Staff in the Prime Minister's office and in the central units of the Cabinet Office realised that this was something that could and should be attempted on a government-wide basis, in order that departments and agencies should be better able to access each other's materials, in order to support more cross-agency collaboration, which has been a major commitment in British government in the second half of the 1990s (6 *et al.*, 1999, 2002; Sullivan and Skelcher, 2002). A specific proposal for the development of, initially, a single 'knowledge pool' for use across government was made in a report from the Prime Minister's internal think tank, reports from which often have the status of government policy (Performance and Innovation Unit, 2000a). Responsibility for development was then given to the Centre for Management and Policy Studies in the Cabinet Office.

As the ambitions grew for the government-wide project, to encompass support for policy makers in joined up or holistic policy making (Strategic Policy Making Team, 1999), it came to be known as the Knowledge Network, reflecting the expansion of aspiration from document management to knowledge management, and eventually that it should cease to be a distinct activity of management at all.[3] Moreover, aspirations for the knowledge pool became increasingly concerned with the aspiration for evidence-based policy making. Pools are defined as vector of three kinds of resources – a group of people supplying and sharing the knowledge; informational resources such as research reviews, evidence banks and the like; and processes such as conferences, workshops, seminars and other online fora.

The Knowledge Network is intended to provide the information technology infrastructure for the pools. Five pilot projects were established to explore the kinds of technological options available, the kinds of change management programme required to enable civil servants and ministers to make the best use of the resource. These were completed, evaluated and reported upon in May 2000. The tender for the implementation of the Knowledge Network was designed to elicit bids from consortia that combined companies with hardware and software technological capabilities with companies that could provide the managerial consultancy support necessary. The winning consortium including telecommunications company Cable and Wireless, the software multinational Lotus, on whose 'Notes[TM]' platform the system would run, although it would be used through browsers, and the information technology company IBM.

The first stage was to develop by the end of October 2000 the network linkages into departmental intranets and networks, with content available initially on a 'read only' basis, in key subject areas. The key priority areas were identified as being a regularly updated directory of all staff with coded indicators of particular areas of expertise, 'basic facts and figures', 'research information', 'media monitoring' and, for a slightly later implementation deadline, a briefing system.

Anyone with access to the GSI will very soon have some level of access to the Knowledge Network, depending upon the level of their authorisation: at the end of 2000, about 55 000 people had access. In principle, some modules within the Knowledge Network could be made available more widely to public servants in any authorities and agencies not on the GSI, either through extranets or through the existing system of portals: there are currently about 500 000 civil servants. Of course, degrees of access will be finely graded. Many documents generated and used in the course of policy formulation are often classified as requiring the highest degree of confidentiality, and even under the UK's new (2000) freedom of information legislation, civil servants' advice to ministers remains protected. It is expected, however, to remain a system to which access will be confined to the executive: there are no plans to provide members of parliament with direct access, although like anyone else, they have access to publicly available material on departmental websites and may be invited on an individual basis to participate in particular extranet supported programmes.

Although the first users were expected to be those preparing briefings on existing policies and initiatives, and initially the main users were press officers, the intention is to develop the system into a support tool for those working on policy formulation and reformulation. It is planned that users will be formally surveyed about their experiences of using the Knowledge Network. Much of the hard information to be made available there will take the form of regionalised facts and figures. In addition, there are online discussions for document archives, and other facilities. Services now available on the Network include a facility for collaborative working among members of the Government Legal Service including a database of government legal forms, a special site for economic reporters working with the Foreign Office, an index of ministerial responsibilities, a database of government IT projects, an electronic library for government constructed from the various departmental libraries' online services with a common enquiries system, a facility serviced by the Treasury for Information sharing on matters affecting public spending. The plan for the system is that it will provide a platform for a new series of knowledge pools in which internal civil service experts, contracted outside experts from consulting firms and academic centres and from business will be able to engage in structured dialogue or looser deliberation using the documents available on the system, not only to exchange ideas but also to work collaboratively on policy projects. Together with the Public Records Office, the system is planned to provide the framework for electronic storage of all government documents by the end of fiscal 2004. More recently, the Strategy Unit has launched a website called 'the policy hub' dedicated to evidence-based practices in policy making.

All departments are expected to consider how they will organise their work flow processes to fit with the aims of the system. The centre has no power to direct them to adopt particular working practices; there are well-known arguments for the view that attempts to impose particular ways of working with information and knowledge can be counter-productive, although there are also contrary arguments to the effect that some measure of standardisation is necessary to get the most from such document and knowledge management systems. Some departments may make their internal knowledge pools accessible through the government-wide system.

A key challenge in any knowledge management system is that of ensuring that people do in fact make all the appropriate documents available, that they make the priority and the time for generating documents to share with others. This requires a particular kind of organisational culture, and as we shall see in Chapters 8 and 9, this is not always straightforward to achieve. The Knowledge Network Change Management Team within the Centre for Management and Policy Studies in the Cabinet Office provides support for programmes of change management in departments to this end. Staff are identified who can provide face-to-face support for users in difficulty. Training sessions are used both to pass on use skills and to try to motivate greater use. Content in mainstream civil service training programmes is to be adapted to develop skills in use, to encourage use, to explain benefits and proselytise for the Network.

Source: Centre for Management and Policy Studies staff, interview with author, November 2000, additional written information from CMPS, March 2001 and from website of the Office of the E-envoy July 2003.

and electronic conferencing style communications within the group, databases of activities, workflow management, group diary and appointment bookings, project management capabilities, as well as extensive search, retrieval and analysis of messages, communications, documents and indices for those documents, and structured databases. In addition, the product provides several levels of security and of data or document access control. Typically, its use requires a client server architecture at the hardware level, and it is often used to greatest effect to sustain group working among dispersed members, often communicating remotely.

Many public authorities use Lotus Notes™ to support group working among staff responsible for administration, services and policy implementation. In the state of Victoria, Australia, for example, there is a strong commitment to cascading its use throughout every department of the organisation, and specifically disciplining all staff to use it rather than to use individual word processing and e-mail systems, and to use it wherever possible in place of proprietary departmental or unit databases (as recommended for government more generally by such commentators as Holtham, 2000).

There are many groups of chief officers and policy staff who do indeed use Lotus Notes™. However, many use it more to support their administration and management. Its principal capabilities that directly support the specific activities of policy making are those of document generation and document flow capabilities to manage the sequencing of circulation of drafts of documents.

Modelling decisions and advising on possible consequences

Decision support systems

Although in both the public and the private sector, the term 'decision support' is widely used, it does not refer to any well-defined or limited set of software functions. Almost anything that could be said to help people make decisions has, sooner or later, been put under this rubric. However, strictly speaking, a true decision support system should provide models with which to handle, analyse and extrapolate from data: a widely accepted definition of decision support systems is that they are 'computer based systems that help decision makers confront ill-structured problems through direct interaction with data and analysis models' (Sprague and Watson, 1996, 6).

The simplest 'decision support' tools are spreadsheet packages that enable users to make alternative projections from a series of data or hypothetical data (in effect, assumptions), and which can then provide a focus for debate. For example, alternative financial projections that can form the basis of alternative scenarios constitute a simple output which can provide some assistance to decision makers. The same simple procedure can be conducted in reverse. For example, if a goal can be defined clearly enough and if the causal relationships between what organisations do and their achievements of their goals are well under-stood and definable – such as those of raising enough revenue in local taxes and central grants to support a target level of expenditure – then fairly simple systems can be used to present alternative combinations of inputs and procedures to achieve that end, or scenarios in which the same elements are combined but at different times. Others focus on using scoring systems to develop 'balanced score card' approaches to appraisal. Many highly sophisticated quantitative decision support tools have been developed as extensions of spreadsheet and statistical database packages. For example, in addition to providing the capability for running advanced statistical analyses upon data sets and hypotheti-cal data about future events or streams, decision support systems can provide expert system modules that will allow complex alternative statements of rules that define causal or legal or logical relationships.

User-friendly front-ends enable decision makers to proceed step-by-guided-step through a prescribed decision procedure. Less structured systems provide prompts and suggestions rather than insisting on a single pathway. More recently developed are neural net based systems that can 'learn' inductively about patterns that can be observed in relationships between data sets and provide additional analyses. Decision support tools are, however, often of the greatest use to the most senior policy makers when their outputs are in graphical form, because bar, pie, line and scattergram charts are often much easier to understand than abstractly presented calculations (Sprague and Watson, 1996).

Expert systems

An expert system is a kind of model, but one which need not use quantitatively expressed relationships. It consists in a set of 'if ..., then ...' rules; some kind of interface that permits information about particular assumptions or cases to be entered, and an inference engine which brings the system of rules to bear upon the case to make some specified kind of inference about it, as requested by the user. Expert systems are widely used in electronic public service provision in those fields where, for example, entitlement to the service is more or less a straightforward and determinate affair, dependent exclusively on the extent to which an individual applicant meets tightly specified conditions, and where there is only administrative discretion insofar as there might be factual dispute about whether those conditions obtain.

The use of expert systems in policy making has been limited. In particular, however, expert systems have been used to develop non-quantitative models of bodies of law. These can be used to test the consistency of those bodies of law, either in draft or after legislative enactment, in order to facilitate revision (Frøkjaer, 1989). Such a consistency testing initiative was conducted in the 1980s, using an expert system model, by the Home Office in the United Kingdom, upon the corpus of immigration and national law and regulations.

Formal models

At one extreme of the spectrum of decision support tools lies the construction, analysis and interpretation of formal models of policy problems, where those models are designed normally to work with quantitative information. This is in fact one of the oldest uses of computers: after the first governmental use in the US Department of Defense contracts to calculate ballistic tables for assorted artillery and missile systems in the 1940s and 1950s, John von Neumann's first serial computer systems are said to have attracted commissions from the US Treasury to run assumptions on very simple models of the US economy. Today, modelling of national economies represents a small industry, and most models contain thousands of equations representing causal and logical relationships between variables. Models are also used in public health policy for understanding epidemiological processes, in a wide range of cash benefit policy fields to understand the relationships between economic and demographic trends on the one hand with public spending on the other, and more recently in several fields of environmental policy to examine the impacts of measures, for example, to curb pollution, upon economic activity and its impact upon environmental variables such as air quality.

There is no doubt that, today, it is hard to imagine certain kinds of policy making being done at all without the use of formal models. The Treasury model of the British economy, for example, is only one of many rival models developed in various parts of government, international bodies, large and small economic consulting houses, financial institutions, academic centres and elsewhere. It would today be almost unthinkable that anyone would contemplate a central bank setting interest rates, a finance minister making changes to the taxation régime, a competition authority making a major decision about authorisation of a takeover or a merger in a very large industry, without at least considering the evidence that can be gleaned from at least several runs on at least one, and preferably more than one model of the economy, using slightly different assumptions and slightly differently deployed policy instruments.

Formal models of policy problems of this kind are expensive and time consuming to build. More than this, they are very data hungry. To have confidence in a model of several hundred equations, one has to be confident about each of the complex and dense web of relationships between all the variables represented there. This in turn requires – at least in the ideal case – that the equations are securely grounded statistical analysis of large data sets collected over some years on a consistent basis, and also on reasonably robust causal models that explain the statistical correlations observed and analysed. Not all policy problems are in this happy condition, even in a research-intensive policy-making culture like that of some departments of the United States federal government such as Energy (see Chapter 6).

A recent report from the Performance and Innovation Unit (now the Prime Minister's Strategy Unit), the long range think tank in the Cabinet

Office, examined the extent and state of the use of formal models of policy problems across central government in the United Kingdom (Performance and Innovation Unit, 2000a). With the new Labour government's publicly stated commitment to 'evidence-based policy making', the issues of the extent, quality, use, relevance, dissemination and infrastructure of support for formal modelling have come to be regarded as important areas of concern. The study conducted case studies of eighteen initiatives across British government using models exhibiting varying degrees of sophistication, which included (among others) models for forecasting road traffic growth, household formation, population growth, long-term care costs, labour market dynamics, tax-benefit interactions, impact of tobacco smuggling, housing rents, waiting times for operations in the National Health Service.

Box 3-d describes an important recent US initiative in the use of formal models.

Box 3-d   Modelling welfare reform in New York State government

The Decision Techtronics Group at State University of New York's campus in the state capital Albany, has undertaken decision support contracts for New York State administrations over many years. Like the British information technology's work on local authority budget setting described in Box 3-f below, the team has developed one speciality in chauffeured decision conferencing. However, it also has great expertise in formal modelling of policy problems, and it has provided state policy makers with an integrated decision support package using formal models of policy problems as part of chauffeured decision conferencing.

Professor John Rohrbaugh, leader of the Decision Techtronics Group describes in a recent (2000) article, the development by the team and the use by state policy makers of a formal computer based model of the dynamics of welfare claiming in the state. Federal and state level politics during the mid-to-late 1990s were often dominated by the debates over welfare reform (Bryner, 1998). Democratic President Clinton's decision to sign the Republican inspired Personal Opportunity and Work Opportunity Reconciliation Act of 1996 was deeply controversial. The legislation ended the previous system of cash benefits, called Aid to Families with Dependent Children, to which lone mothers were the principal entitled group, and created a much less generous scheme, based around welfare-to-work programmes, over which states had much greater discretion. The need quickly to develop affordable, locally politically acceptable schemes, that were consistent with the range of other labour market and social policy programmes to which they were already committed, presented state level policy makers with enormous challenges.

Agreement was reached between state level policy makers and the Decision Techtronics team that the team would build a model of the problem to which the state must design its two schemes, one for the Temporary Assistance for Needy Families programme and for the state Safety Net initiative. The model, like – for example – the British Treasury's model of the British economy, which is now available for anyone to run their preferred tax changes upon to see how the model predicts their results, was built around a huge battery of equations. These equations seek to represent in simplified form the key relationships between, for example, changes in the gross domestic product of the state's economy and the numbers of people flowing into entitlement categories and the periods during which different groups of entrants might be expected to re-enter the labour market with and without support, the relationship between demographic changes in the state's population and numbers at risk of entering entitlement categories, and so on. Fully seven hundred equations were developed, modelling factors influencing both stocks and flows of claimants, grouped by a classification of their needs, and the range of potential outcomes. Flow charting tools enabled policy makers to understand the key relationships used. The model was supplemented with a range of tools by which alternative policy proposals – which could be defined quite finely in terms of small incremental changes to expenditure levels and the distribution of resources between services – could be run against the model, to see what impacts it would predict. Rohrbaugh compares the experience for the policy maker of using the model to that of a game player or a professional pilot using a flight simulator system: alternative manoeuvres can be tried out, with rather lower levels of risk than in an experiment with people's lives and livelihoods at stake.

Policy makers in the governor's office and the executive did indeed make use of the model, to reduce the range of policy options under serious consideration to a handful. Reflecting on the policy makers' experience of using the model, Rohrbaugh writes:

> As in response to any dynamic model, some administrators continue to trust their own intuition and judgment far more than the financial and social projections from the welfare simulator. In fact, because of the complexity of the welfare system with the delays, feedback loops and equilibrium-seeking behaviours that one would expect in such a system, the long-term impacts of some popular policies are starkly counter-intuitive. But the assumptions and projections of the model, unlike the subjectivity of much human judgment, can be thoroughly traced back and critically examined. (Rohrbaugh, 2000, 529)

He reports one politician commenting that

> It will still end up being a matter of leadership and willingness to act on what is learned (Rohrbaugh, 2000, 530)

Source: Rohrbaugh, 2000.

Rohrbaugh and the anonymous New York State politician he quotes are surely right, that, even when this has been done, there remains much of the process of judgment still to be gone through. Just how much, we shall see in Chapters 5 and 7. Indeed, what Rohrbaugh does not stress but which is of central importance, is the role of judgment in the decisions about which policy problems are worth the investment of developing models, the judgment calls involved in deciding which relationships are regarded as either interesting and important enough or sufficiently robustly evidenced, to be worth including, the judgment calls about what counts as the priority assumptions to run against such models, the judgment calls about the sequence in which assumptions are run against the model which can so powerfully influence the ways in which it is used.

None of this is not necessarily a limitation of models like this, but it is a severe limitation upon the power of a certain view about formal modelling of policy problems that has welled up periodically ever since Leibniz' first attempts to develop such models in the eighteenth century. That flawed view is the one that holds that mathematically constrained optimisation of particular variables in such models can uniquely determine the best policy decision, and the authority of the model itself suffices to ensure that policy makers will subordinate all other concerns of political symbolism, partisan or class loyalties, rival rankings of what is important, considerations of political acceptability, considerations of what publics will bear, considerations of what responsibilities politicians can legitimately be expected to shoulder, in order to commit themselves to the straightforward effort to implement the policy design marked as optimal in several runs upon such a model. Of course, no one today would publicly admit that this idea informs their thinking. Yet policy makers who would vehemently deny thinking any such thing when it is stated baldly, can nevertheless behave and bring expectations to bear upon decision making that seem to reflect this assumption. Box 3-e provides a case that should serve as a parable.

Box 3-e   A short cautionary tale about modelling, expectations and policy making

One freelance public policy consultant in Britain worked with a consulting house and an academic centre, to advise the police authority in a northern English city. Using police data of reported crime by district and some local survey data on crime not necessarily reported to the police, the team assembled a small and simple model of a few tens of equations that seemed to perform tolerably well in retrodicting levels of crime, according to those data sources, in different parts of the city. On seeing the quantitative results, the police authority members, having initially been cautious, and having been quite specifically warned of the limitations of the model, became very excited. They talked of using the model's 'prediction' of next year's crime figures, immediately to change the allocation of resources for policing around the city, to allocate priorities differently between different types of crimes, and so on. The consultants and modellers, on the other hand, quickly became alarmed. They pointed out the limitations of the crime data, the lack of robust causal understanding of the relationships, the extreme sensitivity of the relationships described in the equations to changes in the local and national economy, to changing demographics, to changes in the wider national and international patterns of crime, none of which could be assumed to be unchanging. They warned the police authority members that mechanical reliance in policy making upon such a model was dangerous. Police authority members responded to this with a combination of disappointment and bafflement, saying, for example, 'if we can't use the model to predict and allocate, what have we spent all this money on it for?'. Only with the greatest difficulty could the consultants and modellers persuade the police authority members to regard the model as but one input, one consideration to be borne in mind alongside many others, in their decision making.

Source: Interview with author under conditions of agreed anonymity, November 2000.

Chauffeured decision conferencing

The term 'decision conferencing' is used to cover a wide variety of practices in management and policy making. Although I have described what was done with the British all-party group as 'decision support conferencing', its practitioners describe it simply as 'decision conferencing'. In principle, any computer supported collaborative and deliberative process within a group that helps that group to come to a decision, and which does so by enabling them both to engage in reasoned advocacy, to represent their preferences, and to have those preferences subject to some procedure of aggregation or conciliation with a view to coming to closure, will be regarded by someone as decision conferencing. However, when it is used strictly, the term describes a quite specific procedure, long pre-dating computer support, that is the intellectual property of an operations research group at the London School of Economics, headed by Professor Lawrence D. Phillips (Phillips, 1988). The procedure consists a chauffeured workshop, in the sense that the members of the decision-making group do not themselves use keyboard and screen, but one of the two facilitators uses a single terminal to develop the representation or model of the groups' preferences, which is presented and used only in the final session of the workshop (for a description of chauffeured decision conferencing procedure, see Phillips, 1989 and Wood *et al.*, 1998). Box 3-f describes the way in which the procedure was being used by one leading British information systems company until relatively recently and to help groups of senior local authority councillors and chief executives make decisions in the process that leads to the setting of a budget and a rate of council tax.

Box 3-f   Chauffeured decision conferencing to support budget setting in British local authorities

A large and long-established multinational information systems company ran, from one of its offices in a city in the north of England, a small practice that specialises in chauffeured decision conferencing for public authorities. Most of the group's clients were local authorities, although some were police authorities, and they also worked with National Health Service authorities and trusts, and marketed their service quite heavily to central government departments and agencies. Most of the decision conferencing workshops they ran were to support local policy makers – both elected councillors and chief officers, and in many cases, other tiers of management within departments too – in setting budgets, although they used their application of the technique and software for many other kinds of decision-making processes as well. Indeed, between the late 1980s and the end of the 1990s, they worked with about seventy local authorities, which represents about 15 per cent of the total number of authorities in the country. The team eventually left the IT company to set up their own consultancy, continuing their work. In every case, one of the team acts as facilitator for the workshop. Typically, in the first decision conferencing workshop conducted with a group of policy makers, a member of the team will also act as the analyst, using the software and developing the model from the preferences and decisions of the group. However, in some cases where they work with an authority over a period of years, they may sell the authority a licence to use the software, and they may train one of the authority's staff to act as the analyst: this can reduce the cost to the client, calculated over the period for which decision conferencing is to be used.

Chauffeured decision conferencing is something that policy makers need to commit time to, in order to make it a success. They need to set aside time in advance of the two-day workshop, in order to prepare themselves with information about the problem for which a decision is required. In the case of the budget-setting process, of course, they will be doing this in any case. However, the particular structured discipline of the chauffeured decision conference workshop requires that preparation to be done in a structured manner. The workshop itself requires two days. To make the best use of the work done by policy makers in the workshop also requires making time available in the subsequent cycle of meetings to make use of the information generated.

The workshop procedure is strictly defined. A large room is required, in which the participants are seated in a horseshoe, facing the facilitator. There are whiteboards on each wall. The analyst sits behind the group, with the personal computer on which the software is loaded. The ideal size for a workshop group is between eight and twelve, although this team has worked with groups of up to fifteen members. The workshop is split into eight sessions, four on each day. The sessions from the beginning until the lunch break on the second day are concerned with what is called 'evaluation', and the sessions on the final afternoon are devoted to 'analysis'. Suppose the task is that of setting a budget. Initially, overall mission and purpose are agreed: this can often be time consuming. Then main categories of activity are identified by output, and the existing allocation of expenditure between them is laid out. Incremental changes, both modest and radical are proposed, or bid for, and the group is required to agree their preferences, by considering the impact of each change bid for, on a set of criteria that are generated in the workshop. Many of the budget change options to be considered have to be worked up during the period of preparation prior to the workshop, and they should be developed by those with specific responsibility for each problem or output. The criteria must all be derived from the overarching mission and purpose. For example, a cut in each output category of spending might be evaluated on the dimensions of impact on service quality, impact on other outputs, impact on the local economy, likely impact on voters' decisions in the next local elections, and so on. If cardinal numbers can be attached to rankings to evaluations of options against criteria, then these are used: if not, then ordinal numbers reflecting the group's agreed subjective preferences are recorded. The facilitator must next force the group to reconcile their preference rankings between options, by asking them to say by how much is option A worse than option B, when they have ranked them differently on some criterion, and so on. Weights must then be generated and applied to reflect the manner in which the group come to accept consistent use of the criteria, in order to generate a simple percentage score for each option on each criterion.

Entering the weighted scores into the software enables the construction of a model of the groups' agreed and debated preferences. The workshop discipline forces a measure of consistency and weighting to reflect preferences on a single common scale that can be represented numerically. These modelled preferences can then be represented graphically. The graphical representation in budget workshops that is often the focus of most attention is that which represents options by degree of benefit ranking on the vertical axis, against costs on the horizontal axis. The model is presented to the group on the final afternoon's 'analysis' session. Those who find the results surprising can check them back at every stage against the scores which have remained upon the whiteboards in the room. The model enables the group to see the consequences of its preference ranking in some detail. If the most preferred option has to be ruled out for some reason – for example, it would be politically too risky to announce it prior to the next election or bye-election – then it can be blanked off in the model, and the next option considered in more detail. The purpose of the final session is to use the model to go through relationships between options, make any final revisions to scores if this proves necessary, and to see how far it is possible to move to commitment to decisions reflecting the scoring, weighting and the choices made using the model. However, it is always made clear to participants that the real purpose of the workshop is to inform decision making, not to substitute for decision making.

Therefore, exploration of commitment is essentially done in order to test the goodness of fit of the model with the preferences of the group.

The decision conferencing workshop is something that is undertaken on the understanding of strict confidentiality, in order to enable policy makers freely and as candidly as possible to discuss options that may be unpopular, that may be risky to announce publicly for some time, and so on. The model built during the workshop is also strictly confidential to the participants.

Unlike many electronic meeting support systems, in chauffeured decision conferencing, because the real work is done in face-to-face debate and argument, and because only the analyst is using software, there is no anonymity for participants. This may reduce the candour with which people may speak, and may – depending on the group and the facilitator's work – limit the degree to which some participants may be willing to think radically in front of their colleagues. However, it has the advantage of enabling commitment to and ownership of decisions made to be achieved more easily.

The team report significant differences in how different kinds of policy makers operate together, make use of the procedure, and react to the modelling procedure. The least successful are those workshops done with authorities in which inadequate time is made available for chief officers to undertake the preparation work of identifying options for evaluation, mapping and understand existing budgets and the likely impacts of particular choices, and so on. Timing matters: after a major change in legislative framework, there is more demand and people are more willing to think radically about options. The culture of particular organisations also makes a big difference. Where there is a very dominant individual in the group, with whom others are reluctant to be seen to disagree, it may be necessary to ask everyone to write down evaluation scores, rather than to debate them verbally. Police chiefs, they report, are sometimes more reluctant to debate with each other candidly than are local authority chief officers: this, they speculate, may reflect that senior police officers move between departments more than local government chief officers do, and they are aware that they may have to work in or head up a department for which they are advocating cuts in a workshop, and so they often tend to defer to the current chief officer responsible for each department in evaluation.

Sometimes it is hard to get people to think in terms of marginal increments upon which the system works. Officers are much more comfortable debating options for changes within services, while politicians, at least initially, tend to want to make judgments between whole service categories. As one would expect, politicians attach more importance to electoral criteria than to more technical criteria about service quality or long-term impact on local economies, and so on. Moreover, although politicians insist on being given material in advance, the team finds that they frequently do not attach any priority to reading it or doing any preparation of their own.

Getting group cohesion on scores is often the most difficult task for a workshop. As one might expect, a workshop with departmental chief officers being conducted in the presence of a finance director is one in which one can expect a great deal of turf defence behaviour and 'waving the bleeding stumps'. The team find that it is often better to conduct a series of workshops starting with departments, and with subordinate managers, and to bring in politicians at the final stage in the series.

In general, they find, councils can make effective use of chauffeured decision conferencing for the same issue – such as budget setting – for about three years at a stretch, unless the area is subject to great turbulence which produces major changes in preferences. After that, the benefits have been gained, scores tend to settle down, and the discipline may not bring the group to make sufficiently significant changes to the things that they have agreed in previous years and are now committed to, that continuation seems to represent value for money. However, some authorities have used it for three or four years, then used some other method but returned to using the system after a period, when conditions, values and preferences have changed – for example, after a change of political control.

The model produced in the workshop clearly represents a snapshot or cross-sectional view of preferences at a particular point in time. Preferences will change after the workshop, if new information comes to light, if political priorities change, or with the brute force of events. In this sense, the value of the model itself degrades over time. The key to effective use of decision conferencing, then, is the judgment made about the right time in the budget process at which to hold it. It should not be too early in the budget cycle, nor should it be too close to the formal meeting of the full council to set the budget and the Council Tax rate, for if the workshop does identify areas that need further thought and work, that will need to be done properly.

The team recognise that there is sometimes resistance to the idea of chauffeured decision conferencing workshops. Some politicians and chief officers are afraid of being forced to disclose information that they have sought strenuously to keep to themselves about their budgets, and may fear very explicit comparison with others in the structured format of the decision conferencing and preference modelling procedure. Some groups of politicians, where there is great distrust within the group, will resist taking ownership of the preference ranking, and reserve the right to disown it later on. Where relationships between policy-making groups are very competitive, where there are rival factions that will not co-operate, where a single autocratic leader regards other elected members or chief officers as there only to agree with them, where there is insufficient trust between members to share information, where politicians are fearful of being criticised for spending money 'on themselves', or where they will only agree to take part on the understanding that they will blame the procedure and the facilitating team if the results prove unpopular, there are few prospects for successful chauffeured decision conferencing.

Source: Interview with IT company staff under conditions of agreed anonymity, April 2000.

Environments that provide integration and storage

Intranets

An intranet provides a variety of services on an internal network within an organisation, ranging from simple internal e-mail through to supporting document flow management and the storage facilities for banks of policy relevant information. An intranet is distinguished from other facilities available on a network by the common hypermedia-linked structure and the common data format of all material, because it is designed to integrate a range of services through a common access system, namely browser software of the same kind that is used to access the World Wide Web. Although intranets have been installed in most public authorities during the 1990s, it is widely reported that only a minority of authorities are really making full use of their capabilities, and for many, they provide little more than expensive internal mail combined with a means for senior management to communicate some messages to staff. The facilities are rather less frequently used for integrating document flow management, group database maintenance, internal publishing and so on (National Audit Office, 1999). However, as noted above, initiatives such as the UK central government's Knowledge Network programme attempt to make greater use of intranet facilities, using Lotus Notes™ through a browser.

Conclusion: overview of e-governance tools

This chapter has used the classification of e-governance tools offered in Box 2 a, to give descriptions of the principal types of tools and some examples of their use in support of policy work in government.

Although some of these e-governance tools can, at least in principle, be used by any policy maker working alone in any setting, some can only be used in the context of groups that are willing to subject themselves to disciplines of structured working, and to learning certain techniques. In chauffeured decision conferencing and what I have called decision support conferencing, the discipline is exacting, specifying the structure of time, the layout of physical space, the kinds of relationships between policy makers that are required in advance. While videoconferencing or electronic whiteboard tools can be used quite casually, their most effective use is after within a structured setting, often using facilitation.

What we know about the distribution of their use in government is striking. Some are used commonly, frequently and their use is widespread: online consultations and the use of spreadsheet software packages are the best examples. The reasons for this widespread use are not hard to find. These technologies are designed for uses that are but incremental extensions of things that policy makers have long done using paper products that are close analogues. Others are in common use simply because they are cheap, easy to use without training, and require rather limited discipline. In this category, the rapid spread of e-mail is the best example.

Others – and here, formal modelling of policy problems is the best example – have been taken up principally within large, well-endowed central government departments, and are used in a hands-on fashion almost exclusively by professional civil service policy analysts, and indeed, by people working with the disciplinary services such as government economists, government statisticians, government social researchers, but hardly at all by ministers, let alone by select committee members or backbenchers. Indeed, with the exception of finance ministers in the Treasury, ministers rather rarely specify with any exactitude the assumptions and analyses they would like run on such models – a fact which is coyly referred to as 'weak demand' in the recent British government report on their use (Performance and Innovation Unit, 2000a). Whether ministers *should* or even *can* be expected to make hands-on use of such models or to specify their uses and place great reliance upon them in making policy is a question that will be addressed in Chapters 8 and 9.

There remain many e-governance technologies, the take-up of which seems patchy, even haphazard. This might seem surprising. After all, most of the technologies in this category are ones for which uses in policy making easily suggest themselves, and there has been no shortage of advocates for the application of such systems to policy making, even as far back as the 1960s (e.g. Simon, 1977 [1960] and 1996 [1969], esp. ch. 6). Perhaps the simple facts of their recent development and their cost, and the uncertainty of policy makers about the benefits they offer is sufficient to explain the limited take-up.

But it is likely that other and more important issues are at stake. Indeed, throughout the chapter, a number of cautionary points have been made about the dangers of over-reliance – if policy makers do in fact do so – upon recommendations made using such systems, and about the kinds of relationships required of policy makers for their successful use. Chapter 4 is devoted to exploring the possible explanations for the slow take-up, before offering some reasons to think that interest in many of these systems will increase in the next few years.

4
The Forward March of E-governance Halted?

In this chapter, the question is addressed directly, just why have some apparently promising e-governance tools not been adopted more readily in recent years? Answering this question is not otiose, nor of mere antiquarian interest. For the answer one gives will shape how one sees the prospects for greater use of at least some tools in the future.

Five types of explanation have been offered in the past, and the chapter reviews each in turn. I shall argue that whilst all have some merit, the institutional forces are often the important underlying ones, and they are often what lie behind the forces identified in the other styles of explanation. In Chapters 8 and 9, I shall present my own account of the institutional forces at work in shaping what differently situated policy makers can do with e-governance tools, but for the present chapter, it is enough to see the balance of forces.

Finally, the chapter turns to the prospects for greater use of e-governance tools in future, and it suggests that there are reasons to think that there will be more demand, because there are some important institutional forces at work that could significantly push up demand for these tools.

Why has take-up of more advanced e-governance tools been so slow?

The key puzzle is why, if the benefits are so clear and potentially large as the developers of these systems claim and as the writers of the vast number of how-to-do-it books on decision making argue, take-up of the more advanced systems has been so slow, despite many of them being available for some years or even a decade now. Very often, the question of whether to invest in them has simply not appeared on policy makers' agenda – a case of a 'non-decision' (Bachrach and Baratz, 1962). But, if the benefits are so great, even this requires explanation.

It is a moot point whether policy makers and senior managers in the public sector have been significantly slower to adopt these tools than their counterparts in the private sector. Many experts report that the interest among company board members and senior managers in business in decision support tools has often been lower than might have been expected, or at least lower than advocates of such systems would have liked (Professor Clive Holtham, interview with author, November 2000; see also Weick and Meader, 1993). If this slow take-up displays some failing or defect on the part of the humans rather than on the part of the software and hardware, which is greatly contested,[4] then it is far from obvious that the public sector is alone in any such deficit, and therefore it is not clear that the political considerations that shape policy making in the public sector are the source of any problem here.

Very generally, when a class of people such as policy makers do not adopt a set of systems such as e-governance tools, we reach most readily for one or more of the following five basic kinds of explanation (John, 1998):

1. *Information network flow explanations*: Policy makers are not yet aware of the potential uses. The communities that provide policy makers with Information Technology (IT) support – both the networks of IT companies and the networks of internal IT departments within the public sector – are not seen as or do not see themselves as sufficiently interested in policy formulation issues nor aware nor culturally attuned to the e-governance issues to make these a priority (c.f. Marsh and Rhodes, 1992; Marsh, 1998).
2. *Perceived cost/benefit ratio explanations*: The systems are seen by purchasers or indeed by policy makers themselves as disproportionately costly for the benefits offered, not so much in the initial investment, but in the opportunity costs of the time required by their policy maker users in learning the systems, entering data, keeping data up to date, making use of the capabilities of these systems. These explanations do not assume that perceptions of costs and benefits are typically accurately or rationally calculated, or even calculated at all. Perceptions are not necessarily informed, nor do they necessarily reflect understanding, serious consideration or even information of what e-governance systems might offer.
3. *User skill and willingness deficiency explanations*: For a variety of reasons, the decision makers are not or do not see themselves as

sufficiently skilled, willing and suitably culturally organised to make the proper use of these systems.

4. *Design weakness explanations*: The systems are not or else are not perceived to be designed adequately to support real decision processes, but rather have been designed only to support the ideals about decision making processes held by system designers, and these are either unrealistic or incompatible with the ideals of decision makers about their decision making processes – another case of the eternal failure of trying to substitute technical fixes for social solutions to social problems. For example, the systems may assume excessive rationalism in both the goals and the means of decision making, more co-operation and less competition, less desire to avoid shared blame for decisions, or less ambivalence about goals in groups, or may assume either more stably egalitarian or, conversely, more stably hierarchical relationships within decision making groups than can be expected to be sustained in politics and public management.

5. *Institutional explanations*: First, formal rules and informal norms, customs and prior commitments, existing legacy systems of decision making in policy making limit the ability and willingness of policy makers to recognise and accept the potential benefits of e-governance systems, lead them to recognise and give weight to costs, weaknesses of design, and so on. Second, when policy makers do invest in and use e-governance systems, these institutional forces limit their interest to those which support established ways of making decisions. In some fields of policy, such as economic policy making, for example, they push policy makers towards e-governance tools that give particular importance to the analysis of quantitative data rather than qualitative tools for problem structuring and creativity, whereas in other fields such as immigration or social security, institutional pressures may lead to the use of expert systems for testing the consistency and completeness of bodies of detailed regulation rather than idea generation or meeting management tools.

I shall argue each of these views has something right about it, but still there is something missing. Indeed, policy makers have not yet understood what is possible. True, the industry has not yet really learned adequately to provide what policy makers want. True, sometimes IT companies often still seem more interested in offering what they think policy makers ought to want. Indeed, at least some and perhaps many policy makers are for various reasons often reluctant to use older and simpler e-governance systems that they have learned how to use but which have limitations. None of these factors will disappear. However, it is also true that the environments in which policy makers operate, form and break groups, make decisions, are themselves changing. Some of the changes that we can expect over the next few years might well create situations in which certain kinds of decision support system would be more appreciated than they have been so far. But the kinds of system that are likely to be demanded are not necessarily the ones that the industries and disciplines of group decision support systems, knowledge management and computer supported collaborative working already have on the shelves. The industry cannot simply wait for events to deliver its increased demand for the kinds of products that it already knows how to produce.

Review of explanations

Information network flow explanations

It is probably true that awareness of more recently developed and sophisticated e-governance tools is rather limited among politicians and salaried permanent policy making staff. However, it is far from clear that even if awareness of what the knowledge management, computer supported collaborative working and group decision support systems industries can offer today were greater, this alone would have a major impact upon demand. For example, in North American business communities, there is probably much greater awareness of what is available and of course greater ability to raise capital and to make investments that may have to be written off than the public sector can hope for, but in practice, demand still languishes below what the e-governance software industry had hoped (Kling, 1997). Other factors seem to be at work.

In a recent review of the development of new uses of information and communication technologies in British local government, one leading commentator offers a professional network explanation of this limited awareness, developed from his observation that electronic service provision initiatives are much further advanced than are e-democracy or e-governance programmes (Pratchett, 1999). Certainly, there is some evidence to suggest that this argument is correct that the local government information technology department communities are indeed, by comparison with other central support departments, relatively isolated from policy makers, better networked with each other than within their own authorities, closer to suppliers in outlook than to their government colleagues, a rather closed group, with values of their own that are more about technology use than about the kinds of strategic issues that concern politicians, and that these are characteristics that are more consonant with the application of their skills to electronic service provision. There is also plenty of evidence that many systems purchasing decisions in the public sector are still being taken principally on criteria accepted by information technology departments, rather than by users, which shows clear evidence of their significance (Bugler and Bretschneider, 1993).

However, it is not likely that this is the whole explanation, and on its own, it is not very satisfying.

First, in this explanation, the role of policy makers who might use e-governance systems is eclipsed by the role of technological specialists, and this runs counter to much of the empirical evidence about the importance of user decisions in adoption, diffusion, innovation in the course of use, for all information technologies (Bugler and Bretschneider, 1993) and indeed for technologies generally. Certainly, the most senior politicians and policy makers are least likely to be first adopters and innovators (Caudle, 1990). British Prime Minister Blair's much vaunted late learning of basic use of a personal computer is a case in point. In general, this argument is a social network variant of supply push explanations for technology adoption and non-adoption, which have widely been found to be of limited power (Bugler and Bretschneider, 1993).

Second, the low levels of demand for more sophisticated e-governance tools is not confined to local government. In the private sector, one leading US commentator has argued (Kling, 1997), demand and use remains lower than many in the supply industries had hoped. In central government in Britain too, demand and usage remains low, and there one might argue that the IT support community is much better integrated into policy networks than it is at local level.

Third, in general, information network flow forms of explanation are unsatisfying. For these are usually only proximate factors. If other conditions are in place for a certain kind of change, achieving reconfiguration of networks in formal organisations (although not in wider publics) is usually not the most difficult of problems to solve. At most, network-level explanations are a necessary part of the explanation, but they are not sufficient (Dowding, 1995; John, 1998: ch. 4).

Perceived cost/benefit ratio explanations

The promise of e-governance systems is that policy makers will gain greater control over the resources they control through the use of such systems, and some studies do conclude that policy makers' and public managers' perceptions about the balance of these benefits relative to their perceptions of costs are the key determinants of their interests in such systems (Bugler and Bretschneider, 1993; Kraemer and Dedrick, 1997).

There are a variety of perceived costs associated with such systems. In particular, there are costs associated with the creation, introduction, development and implementation of systems.

Almost all e-governance tools require the making explicit and the formalisation of information and knowledge that was previously implicit. This is time consuming, not only in data entry but in the intellectual labour time of capturing, formulating and standardising knowledge into the categories agreed and then setting it up for the system to be used; this is still true even if some of the work of data entry in some systems is devolved to citizens entering their own details (Snellen, 1989, 14ff.). Maintaining knowledge bases up to date and in forms that remain useful for the changing problems that policy makers face is a time and resource consuming task.

Of course, in any organisation, getting knowledge out of the heads of staff and into an explicit, standardised, easily communicable form is time consuming and difficult. Indeed, often the effort by senior management to do this will create conflicts because unless major efforts are made, highly skilled and knowledgeable staff typically have too few incentives to want to reduce their uniqueness, their indispensability and therefore their relative advantage over others by making their knowledge available (Buckingham Shum, 1998). In practice, many groups using decision support systems find that they need to allocate time to activities other than sustaining organisational memory, and that the opportunity costs are simply too great.

Second, there are costs associated with use. Some of these are costs to organisations or to public authorities as a whole, including the costs of maintenance and preparation. For example, systems designed to support group decision making in meetings can be very resource intensive because they require a great deal of prior preparation for meetings and structuring of decisions (Watson *et al.*, 1994). Other costs are private costs to particular managers or departments. For example, many systems are introduced precisely in order to support more integration or holistic working (6, 1997a; 6 *et al.*, 1999, 2002), and perceptions of the costs to departmental policy makers' autonomy can be important barriers to their acceptance and use of new information systems to support policy work (Kraemer and Dedrick, 1997).

Moreover, the benefits of using more sophisticated e-governance tools do not show up immediately. Typically, it is very difficult to point to particular decisions that clearly would have been made both differently and with clearly worse consequences without the use of an e-governance system. Rather, if there are benefits, they are to be expected to show up over periods of time, in the overall pattern of decision making, in the kinds of questions that policy makers are able and willing to ask of themselves and of others, perhaps in the confidence that decision makers feel about the basis of their decision making, and in the medium- to long-term increased capabilities for defending decisions made at the point when they are held accountable, whether that is in public in front of the media or before scrutiny bodies such as audit agencies, select committees or ombudspeople.

Many individual decision makers may feel that their own tenure in their posts is unlikely to be long enough for them personally to experience much benefit from investment in these systems within their current departments, agencies, or authorities. However, there is nothing specific to governmental policy making contexts about these problems. They can be found in every sector, and to the extent that they can be resolved, their resolution depends upon the ways in which senior politicians, senior managers and others can institutionalise incentives for decision makers at every level to operate with the long term in mind, and to align short-term perceptions of personal benefit with long-term aspirations for organisational gain.

It seems likely that perceived cost/benefit ratios may explain significantly more of the low take-up of specialist e-governance tools than information network flows. However, these are often principally proximate factors, masking the work of deeper, underlying forces.

User skill and willingness deficiency explanations

There are several kinds of user skill and willingness explanations, and each must be discussed separately.

The first is concerned with the abilities of public sector organisations to purchase and implement information systems of any kind, whether for services, transaction processing or e-governance. It is often argued that the restrictions on investment required in the public sector for financial accountability impose rigidities, that purchasing systems lead to long lead times in implementation by the end of which the system introduced is out of date, that purchasing rules about accepting lowest tender achieves false short-term economy at the expense of either efficiency, value for money or effectiveness, that the limited span of control that public sector project managers can be given within civil service rules means that project oversight and control are not adequately matched and therefore, the total take-up of new information systems including e-governance systems is much lower than in the private sector and also lagged in time in the adoption of innovations, by comparison with commercial innovation. Many public sector reformers would deny that these are inherent features of public organisation, and propose reforms to introduce a much more entrepreneurial bias across the public sector and a correlative relaxation of concern about probity where that is the source of caution and internal over-regulation (most famously, Osborne and Gaebler, 1991). It is perhaps too early to tell whether the fatalists or the optimists in this debate are right. However, there are enough examples of more flexible public purchasing and investment from around the world to cast doubt on the more deterministic position that government is inherently too rigid to do e-governance well.

The more advanced e-governance tools are probably seen by many users as more demanding to use than many routine office applications. Whether these tools are intrinsically more difficult to use, because of what they do, or because of the particular ways in which they are designed, or whether the perceived difficulty is really a matter of famil-iarity, is not yet wholly clear. For example, if a now routine office appli-cation such as Microsoft's Outlook XP^{TM} – a tool for integrating faxes, e-mails, notes, task lists, diaries and calendars, activity record, with a database of contacts, that can be used as general office working envi-ronment from which to launch and integrate applications – were to have been presented in a form as fully developed as it is in the Microsoft Office XP^{TM} suite, to lay users in, say, 1987, before graphical user inter-faces were in universal use and before the integration of external com-munications with task planning and diary management was conceived as a routine bureau operation, no doubt it would have been thought dif-ficult to use. Indeed, someone who made full use of all the functionality of Outlook XP^{TM} as the centre of their office work would use actual working practices that are quite different from those who do not or from the practices she herself might have used prior to using Outlook XP^{TM}. Perhaps the same is true for many sophisticated e-governance tools.

On the other hand, user skill deficit explanations can be turned around. It can readily be argued that users should not need to be partic-ularly additionally skilled to make good use of a well-designed system that really meets their needs, and that ease of use considerations are what really lies behind skill deficit claims. Even intrinsically demanding tools can be made easier to use through intelligent and sensitive interface design. However, it has been a long-standing criticism of many group decision support systems that the ease-of-use issues have not been thought about carefully enough, that they tend to assume that users think the way that software designers do, that data capture is cumbersome and time consuming, that the basic concepts by which users are expected to organise their knowledge or contributions to argument are far from intuitive, and that these systems make unreasonable demands about the motivations of users (e.g. Bannon, 1997). In self-criticism, some of the leading innovators in the field have accepted that some computer supported collaborative argumentation systems have been too demanding to use (Conklin, 1999).

Probably more difficult to assess is the question whether policy makers in government are less willing to use such tools than executives on comparable positions of authority and responsibility for strategy development in other sectors. There are some suggestions that this might be the case.

First, there are difficulties with public acceptability. There are cases where many local authorities have delayed even providing elected councillors with personal computers and e-mail facilities, or access to council intranets, on the decision of those elected members themselves. The reason is simply that elected politicians sometimes fear that being seen to spend taxpayers' money 'on themselves' will be unacceptable when the public is concerned to see all available resources spent on frontline service provision, rather than on 'administration' (Ranerup, 1999). Politicians are understandably reluctant to make the case to the public or investment in management systems and greater capacity for appropriate priority-setting and oversight, despite the fact that this is what they are elected to do, and despite the popular demand for political oversight and accountability of public services. When public trust in politicians is low, to argue publicly that there is no medium term zero-sum relationship between governance and service provision seems a risky thing for a politician to do. By contrast, shareholders are less likely to balk at investment in the capabilities of the board to develop strategy, make effective use of meetings, exercise scrutiny over management and so on.

Second, despite the wider social trends towards more use of computer-mediated communication, one of the activities that politicians must engage in, and which represents a centre of gravity for their work, remains determinedly face-to-face – namely, political communication with citizens. For all the investment put by political parties in their websites, the high profile forms of political communication remain walkabouts, battle buses, chartered aircraft, speeches from soap boxes, tanks and lorries, pressing the flesh and kissing babies. 'Surgeries' in person in the constituency – drop in advice sessions in which constituents can ask for the politician's help with problems they have in their dealings with public authorities – remain an essential commitment of ward councillors and national politicians alike. Face-to-face conversation with strangers remains the basic skill of the aspiring politician. Face-to-face communication capabilities rather than analytical or decision making skills are vital in securing or at least shoring up flagging public trust. It can be argued that this colours everything that politicians do, far beyond the requirements of communication with citizens. It could be argued that, at least for many if not all of them, the same style spills over to their work as ministers, as leaders, chairs and mayors. One of the *deformations professionelles* of the politician is, it might be argued, a reluctance to engage in communication by other means. Indeed, even in one of the US local authorities most committed to developing e-government, Fairfax County (to be discussed in more detail in Chapter 6), elected politicians still insist on receiving printed copies of agenda papers for their meetings.

Third – at least in a country like Britain with its electoral rules, accountability structures and party cultures geared toward political leadership and strong majorities relatively autonomous from other interests in the wider society (the same may not be true in more consociational democracies) – politicians see themselves as in control of decision making. At least during the period when one familiarises oneself with a group decision support system, one feels that one is surrendering a measure of control to the technocrats who train, support, modify and develop the system. Perhaps politicians, who feel the breath of the public on their necks more than other managers, are more reluctant to accept this situation. Politicians are people with strong views and ruthless ambition. To see themselves, even for a period, as a policy learner rather than a policy teacher, is something they may not be willing to accept with even such humility as commercial executives may muster.

Fourth, it is a common experience for those making presentations to politicians, at least in Britain and in culturally similar countries, that they dislike graphical representation of ideas more than people of similar seniority in other walks of life. Tools that require them to think in this way cut against the grain of the culture that values the debating rhetoric of the chamber or the speech to major business or trade union groups, the face-to-camera soundbite of television, the rough banter of backroom deal-making within the parties.

I do not know of any major study of the validity of these arguments, let alone one that compared their power across countries with different political cultures. However, if any one lesson comes out relentlessly from all literature on computer supported co-operative working, management information systems, computer assisted argumentation, and so on, then it is that introducing technology alone does not and cannot change the cultures of users. If users are to work in different ways, change their expectations, they need to see benefits from doing so that are more substantial than being able to make use of a computer-based system (Bannon, 1997; Kling, 1997; Conklin, 1999). For example, they need to see enhancements to their scope of action (Berry *et al.*, 1998), increased status within the hierarchy of decision makers as a result of using such systems (Nedovic-Budic and Godschalk, 1996; Nidumolu *et al.*, 1996), at the very least no increase in the degree to which the system provides surveillance capacities over their work for superior decision makers (Overman and Loraine, 1994; Hayes and Walsham, 2000). Indeed, these studies show that even in the private sector, as we might expect, personal benefits are more important than organisational ones. Organisations introducing such systems need to do a great deal of work on the social and cultural factors before they can expect motivation, active use, and integration of such systems into decision making (Finlay and Forghani, 1998). For example, some meeting management systems for decision making really require facilitators (Niederman *et al.*, 1996), but this calls for an acceptance of at least temporary subordination that is not always easily accepted, particularly by some elected politicians (Kling, 1997). Finally, e-governance systems will only be successful if they are closely coupled with already well-developed underlying management information systems (Karagiannis *et al.*, 1994).

This suggests that if there are benefits to be gained from the use in government of advanced e-governance systems, then a great deal of face-to-face work (Dennis, 1996) with support from party leaders and senior civil servants, mayors and leaders and chief executives, will be needed to cultivate the norms, manage the expectations, and sustain use among politicians and policy makers. A combination of emotional detachment from setbacks, enthusiasm for gains and long-term focus by top political and officer management is necessary (Newman and Sabherwal, 1996). It is crucial that e-governance systems are not framed as technocratic tools that gloss over conflicts, but as conflict management tools (Jones, 1994).

We may conclude, then, that while skill deficits are unlikely to play a major part in explaining low take-up of more sophisticated e-governance tools, problems of willingness may be important, and that these problems would need to be addressed very directly in any major e-governance initiative. But, one of the major willingness problems concerned public rather than policy maker perceptions of acceptable cost. Finally, as we shall see in the next section, even the problems of limited willingness may have as much to do with design weaknesses as with the cultures of politicians and senior civil servants.

Design weakness explanations

It has been a long standing criticism of many group decision support systems that they are overly 'rationalistic', that they fail to match the real character of decision making in organisations generally, and in politics in particular (Jones, 1994; Bannon, 1997), and that even if policy makers tried to be technically rational, it is not the case that enhanced information and knowledge management in fact produces any greater control (Overman and Loraine, 1994). By 'rationalism', 'technical rationality' or 'policy rationality' here, I mean a narrowly technocratic rationality, in which policy efficacy and cost-effectiveness are all, in which the true goals of policy action or the ones publicly avowed, which are 'internal' to policy design, and from which such 'external' goals as partisan advantage, career advantage for particular politicians, defeating rivals, improving public profile for an administration, and so on, have all been excluded. The point is certainly not that politics is not rational: indeed, within the institutional constraints, it is highly instrumentally rational for its particular goals, as Thucydides and Machiavelli were well aware. But this is a much broader rationality, measured by the range of goals encompassed, the relaxed constraints, the range of available means, than of the policy technocrat.

In the field of legal interpretation – which, of course, at least in the appellate courts, often has major policy implications – the same criticism has been made of expert systems' excessive rationalism, literalism, misleading by failure to model context, inability to handle counterfactual conditions, weak handling of interactions between bodies of law including conflicts of law and uncertainty between national jurisdictions (Leith, 1986; Frøkjaer, 1989). While neural net systems may, within constraints, be able to model decision making in a way that is less rationalistic and more like real policy making, their limited ability to display chains of reasoning back to principles or the mode of application or interpretation of principles limits their value. These arguments have a long pedigree in the criticism of the aspirations of artificial intelligence (for critical discussions see Weizenbaum, 1984 [1976]; Dreyfus and Dreyfus, 1986; Whitby, 1996; 6, 1999c, 2001a,b).

Rationalistic models in political decision making have been an aspiration for at least some people for hundreds of years, but the high tide of technocratic rationalism in politics came in the post-war years. That what policy makers actually do has never met those standards is the oldest news in the discipline of political science. The key remaining disagreement is between

a. those who think that political decision makers do not try to conduct decision making in order to make rationally better policy for specific policy goals but rather pursue short-term electoral, career, partisan, ideological, group-hostile, symbolic, fundraising, social status, ritual or other goals for which policy is merely the arena, the occasion and the foil, and therefore, information systems play a ritual, symbolic and rhetorical role (Nidumolu *et al.*, 1996), in which perhaps the more naïve policy makers sustain their fondly cherished myth that their decision making makes some difference; and
b. those who think that within some constraints of what is cognitively possible and politically acceptable, policy makers do in general actually care about policy effectiveness and outcomes and try, within limits, to make at least incrementally 'better' decisions, even if there is scope for large disagreements about standards of improvement that might count as 'bettering' policy making.

At the most basic level, the goal of most knowledge management and group decision support systems is to enable group members and other stakeholders to understand one another better, to model and make palpable to others, the mental representations of problems and putative solutions of at least one group of decision makers. But on the first view, politics functions, on the contrary, by the organisation of misunderstanding and mutual unintelligibility, rather than by the promotion of mutual comprehension – for in situations of mutual incomprehension, power can be exerted much more effectively and legitimately. Even within the narrow top decision making group such as the Cabinet, mutual understanding is risky, for these are people who have most to fear from each other.

It is not possible to test these rival theories here. In fact, it is hard to know what would be accepted as a test that would settle the matter, since almost any conceivable piece of evidence can be interpreted, given suitable definitions of 'symbolic' 'ideological', and other such, to support either view.

The first issue that does matter for the present purpose is whether e-governance tools can be developed and offered that meet the interests and requirements of policy makers as they are, whatever one's view of the true nature of those requirements is. The second is whether the requirements for e-governance tools of elected politicians and the civil servants or local government officers who serve them can be entirely compatible. For if one takes the view of civil service neutrality that civil service policy advice should be more technocratic, more focused on the 'internal' goals (6 and Forder, 1996) of solving policy problems than the policy commitments of politicians need be, then questions arise very quickly about how the two can use the same systems together, how far politicians will be willing to vote public funds for systems that at the very least may work on different rationales from their own approach to policy problems, and so on.

There remain in many decision support systems, many commentators argue, design weaknesses of excessive rationalism, and unwarranted assumptions of greater co-operation and shared team values among decision makers, and these may indeed play some part in explaining low levels of demand hitherto. But they probably do not explain very much of the problem. For the rationalism implicit in many models and simulations can in fact readily be turned to the account of policy makers looking for legitimation of their policy making with the public. Even tools that do not (unlike models and agents) generate evidence that can be presented in public can be used for this purpose to some degree.

Institutional explanations

Certainly, there is good reason to think that institutions in general are among the most powerful shapers of perceptions of costs, benefits, opportunities and risks (Douglas, 1982a, 1986; March and Olsen, 1989; Immergut, 1992; Weaver and Rockman, 1993; Peters, 1999; 6, 2003).

A problem with many kinds of institutional explanation is that it is not easy to identify just which institutions are at work, and another is to be sure that institutions are genuinely different from ideas or 'cultures'. Are the relevant institutions that shape perceptions always the formal and informal ones that operate in the immediate working environments of the decision makers such as office hierarchies, existing legacy computer equipment, practices and norms of consultation? Or are the important institutions the overarching ones of the general nature of decision making – presidential versus parliamentary systems, the legal authority of the executive, the formal powers of delegation under standing orders, and so on? Or are the institutional forces that shaped those individuals in previous working contexts or even during their primary socialisation more important? Institutionalist theories have been advanced that privilege each of these in general terms, but presumably all of these are relevant in some measure.

One kind of institutional explanation for trends in government use of information and communication technologies is the idea of an organisation's information domain (Bellamy and Taylor, 1998), or the kinds of information that an organisation has institutionalised its capacities to detect (Hood, 1983) and process, information about what is going on in its environment and in the fields with which it must deal. Recently, some researchers working on governmental use of information technologies have attempted more synthetic approaches, recognising the roles of a wide range of types of institutions (Fountain, 2001).

However, there are many kinds of institutional explanation, and although there are good reasons to think that institutional forces lie behind many of the issues stressed by the other kinds of explanation, the hard work of explaining the institutional forces that shape the use of e-governance tools remains to be done. This heavy lifting is the task of Chapters 8 and 9.

On balance, then, I suspect that the least important factor – taken alone – is information network flow and the single most significant explanatory factor is probably perception of the ratio of cost to benefit; however, these perceptions of costs and benefits are at least partly shaped, although not necessarily so conservatively as some institutionalist accounts suggest. Information network flow may be in part a derived factor from institutional forces. All these explanatory factors have some role, they interact and work together: a credible explanation will have to be drawn upon several factors, and the real question concerns the relative weight of those factors in the overall balance (John, 1998).

The prospects for greater use of e-governance tools in the near future

Despite these forces limiting take-up, there are some reasons for thinking that the present levels of governmental interest – at least in Britain and similarly placed countries – in acquiring and using more advanced e-governance tools may rise significantly in the next few years.

It is possible that the movement, now worldwide, towards more open government, will have impacts on the demand for sophisticated e-governance tools. In principle, openness increases accountability, and opens up to greater potential scrutiny, the ways in which decisions are made. Accountability should be a spur to innovation in decision making and to the desire for greater capability for demonstrating the sound basis on which decisions are made. One way to do this might be to use knowledge management systems to capture and support organisational memory, to use document flow profiling systems to show that procedures have been followed and so on.

It is true that this may be a long-term pressure, and that in the short run, openness may have the reverse effect. For example, when President Clinton came into the White House, he brought with him a group of staff members for whom internal e-mail was the obvious and appropriate way to communicate. Initially, they conducted a great deal of policy discussion by internal e-mail. However, when they realised that all internal e-mails were legally required to be stored and could be made available to congressional enquiries, the practice declined. E-mail began to be used mainly to arrange face-to-face meetings instead (former White House staffer, personal communication, from seminar under Chatham House rules). In the same way, it is sometimes argued that forcing open disclosure of civil servants' policy advice to ministers will lead to fewer pieces of advice being committed to paper and more purely oral presentations.

Indeed, openness can be ambivalent in the incentives it creates for use of e-governance tools. However, there is no reason to believe that the inhibiting effect will always dominate. Just as the pressure of accountability and openness has led the Treasury and the Bank of England to greater use and public availability of their computerised modelling tools, the same effect can be expected to be significant in other fields for other e-governance tools.

In much of the developed world, a new phase in reform of government organisation had begun by the mid-1990s, which made some decisive moves away from the agenda of the 'reinvention' or 'new public management' commitments that dominated the 1980s and the first few years of the 1990s (6 *et al.*, 2002). In the current phase, the prior emphasis on the dedicated agency, defined functionally, has been radically re-cast, and the emphasis is now on integration, holistic working or what British Prime Minister Blair, called 'joined up government'. This requires much greater effort in collaboration, above all in policy making, between ministers, departments, agencies, and territorial tiers of governments (6, 1997a; 6 *et al.*, 1999, 2002). Indeed, it calls for collaboration at various possible levels with people often and typically at some distance. It is worth thinking about how far we can expect e-governance tools to assist in this process.

Joint policy work is not straightforward. Certainly, there is no reason to suppose that it can be done effectively without face-to-face meetings. Systems that facilitate communication at a distance, from the simple telephone call to the most sophisticated online group decision support system, are typically most useful in furthering collaboration and intensive joint work when trust, commitment, understanding and attention to jointly defined problems have already been developed and settled face-to-face. And it is not typically true that people can easily and efficiently work together for long periods across distances without renewing those basic underpinnings of their relationship through face-to-face meeting from time to time. In general, for example, research on decision making groups' use of videoconferencing systems suggests that they are most useful for updating, information exchange, handling queries arising from text information exchange, but not for team building, brainstorming, coalition formation, negotiation, or complex problem solving (Mosier and Tammaro, 1994).

However, there is scope for much greater and more intensive use of a variety of communication tools in government to conduct collaboration between face-to-face meetings. Indeed, some group decision support tools are often best used away from the face-to-face setting. Contrary to the fond belief of many managers, group face-to-face brainstorm sessions are generally found by organisational studies to be a poor way to generate ideas, but electronic tools for this that allow asynchronous and synchronous multiple dialogue streams have been found to be more valuable in recent studies (Dennis et al., 1997).

Holistic working often requires more lateral working and collaboration between people with different professional and organisational accountabilities than did the public management of the reinvention era, both at around the relationships between policy problems and around possible linkages between the varieties of intervention available. Some problem structuring tools lend themselves well to this kind of joint policy work (Massetti, 1998).

However, one of the key issues for e-governance support for holistic working is the development of banks of policy relevant information across agencies, accessible over linked intranet systems. At present, there remains at least in the British case, some way to go in this respect. One problem for joint working between different tiers of government is the fact that local authorities are not granted routine access to the Government Secure Intranet (GSI) system. Even security permissions are not granted consistently to access content across departments and agencies within GSI; hypermedia connections between departmental information banks over GSI are not well developed, thus limiting the usefulness of the system for fully joint policy analysis.

However, the availability and even the purchase of technology to facilitate collaboration alone cannot and will not bring about willingness to collaborate. Fountain (2001) has documented in the United States, the history of efforts during the Clinton presidency to introduce an integrated system for tracking trade flows, which would eliminate the duplication of requests for information by customs authorities, transport regulations, health and safety regulators and others, under the rubric of an International Trade Data System. Achieving an integrated system required commitment to joint working that was simply not present: Fountain finds that the US Customs department effectively retained a veto over the project, despite the commitment of Vice-President Gore and his central team.

Institutional perspectives suggest that only when there is prior organisational as well as senior political commitment to collaboration can the possibilities be realised for the use of information technologies to support holistic working (6 *et al.*, 2002).

As I have argued elsewhere, a key part of the response by national governments to the transnationalisation of economic life and other issues has been innovation across a spectrum of styles of cross-national collaboration in regulation (6, 2002b).

Some national regulators are already pursuing these strategies by sharing information across electronic networks. One recent study suggested that in the European Union alone, there were by 1996 already some thirty public sector transnational electronic information systems in use (Kroon, 1997 cited in Cavaye, 1998). E-mail listerv arrangements link national data protection regulators through their Standing Conference, enabling them on a daily basis to pool information about potential conflicts of laws, national priorities, regulated transnational companies, and so on. In the more developed field of pharmaceutical product regulation, joint working is much better developed (Vogel, 1998).

Finally, the development of the electronic provision of public services and of electronic democracy will probably itself provide a spur to the use of e-governance systems. For the automatic capture of data – much of it, for reasons of data protection, in anonymised form – from electronic interaction with citizens and businesses will provide an invaluable data resource for analysis and policy oversight, to which scrutiny committees such as the House of Commons select committees and the audit agencies will want to have direct online access to, in the course of their deliberations, and which many legislators are likely to want to have direct access to, rather than having to wait for some time for Parliamentary Questions to be answered to provide the analyses wanted. The availability of these data and analyses could well become part of the common culture of policy working, and will feed into the policy formulation process, just as it has begun to in the boardroom in many businesses.

Similarly, while online democracy and public consultation lies outside the scope of the present study (see Hague and Loader, 1999; Wilhelm, 2000; Kamarck and Nye, 2002), it is quite possible that as public consultation over electronic networks becomes more common, it may have cultural spillovers to the ways in which policy makers within legislatures and the civil service, local councils and their officer policy staffs, all conduct their policy work.

These are at most drivers for possibilities. None will secure that any particular tools will achieve widespread adoption, nor are they sufficient to guarantee that using e-governance tools will become the daily work of politicians, chief officers and advisors. Rather, these factors provide spurs for greater use of information and communications technologies in policy work, leaving the content of that work more powerfully shaped by the institutions of the policy process.

Conclusion

The basic drivers of the take-up of e-governance tools are institutional; these forces really explain what lies behind the other things to which we turn to understand who uses what tools. As we shall see in Chapters 8 and 9, institutional forces do much more than this: for, rightly understood, they help us to see the constraints, the opportunities and the conflicts that structure the policy process. This study is concerned more with the informal institutions that shape the work that policy makers do than with the formal institutions of the constitution or particular well-established organisations, except insofar as these formal institutions do their work through the effects of informal institutions. On the other hand, changing formal institutions plays an important role in the argument that there are reasonable prospects for growing demand from policy makers for digital support. Again, however, these shaping factors will work, if the overall argument of this book is accepted, to the extent that they have impacts upon the informal institutions that govern the relationships between policy makers and the ways in which they use information, learn and exercise judgment.

If the argument of this chapter is accepted, that there are both understandable reasons that explain the slow take-up of many advanced e-governance tools, but also to expect rising adoption in the near future, then we need to consider what the consequences of this changing trend might be. This is the task for Chapter 5.

5
What Difference could E-governance make? Fears, Aspirations and Theories

What might be the consequences of significantly increased take-up of more sophisticated and recently developed e-governance systems across government? Will these tools sustain 'better' judgment and governance, by the standards and interests of any of the principal interests that shape policy processes? Is it meaningful even to ask whether they could 'better' support anything recognisable as 'the public interest'? Broadly, there are four rival answers to this question on offer. Each can be read as one possible future scenario for understanding the potential impact of e-governance.

Rationalisation

First, there are those who argue uncompromisingly that the use of these technologies represents a major once-for-all improvement in the capabilities of governance, the quality of judgment and in at least the possibility of greater technical or policy rationality in political decision making (Stevens and McGowan, 1985, 177–83; more recently, Tapscott, 1997a,b; Cloete, 2001a,b,c). On this view, the only price is the cost of investment and some running costs. Indeed, the use of these systems will steadily reduce the costs to policy makers of acquiring, ordering, coding, organising, selecting, managing and using information. Therefore, the systems will more than repay their initial costs over their life (argued by, e.g. Reschenthaler and Thompson, 1996). This optimistic view is based on classical cybernetic theory (Wiener, 1948). It holds that that (and here, I follow Overman and Loraine's 1994 useful summary), information decreases uncertainty, slows entropy, increases control (decreases variance) by feedback and deviation correction, and in general that more information enables more control (or at least up to a point that we have not yet reached, if the relationship between information and control is conceived as curvilinear). Increasing control is supposed, on this theory, at least under conditions of liberal constitutionalism, to lead to a greater emphasis on technical rationality and policy efficacy, and to decreasing focus on political considerations.

It is relatively straightforward to obtain evidence for the control theory if one limits oneself to asking no-longer-new, now familiarised decision-making users about their satisfaction with a particular technology, as studies of e-governance in Norwegian (Ytterstad and Watson, 1996) and Australian (Hasan and Hasan, 1997) local government have done. However, people tend to engage in rationalisation and cognitive dissonance reduction, and so they tend to report that the technology provides them with valuable information and with greater control. A leading Dutch commentator on e-governance there presents some qualitative evidence for reduced incrementalism in budget-setting arising from the introduction of spreadsheets and more advanced financial planning tools (Van de Donk, 1998), but the thesis still lacks the comparative quantitative support it would need. Such subjective evidence needs to be treated with great caution, for all the reasons now well-established in the studies on limited rationality in the policy process.

Finally, there simply is no evidence that the use of e-governance tools leads to the depoliticisation of politics. On the contrary, even the few cases presented in Chapter 3 show clearly that political considerations invariably shape decisions to use e-governance tools, the selection of particular systems, the manner of use and the uses to which the results are put.

The price of reason

A second cluster of answers to the question accepts at least the possibility of greater control, quality and rationality in decision making, but insists that this comes at a price. These accounts argue that safeguards are needed, lest that price be too great in terms of

- loss of citizens' individual liberty and privacy, from the ways in which personal information tends to be used and abused by policy makers as they match data from many sources (Raab, 1997);
- erosion of citizens' collective (democratic) influence over governmental decisions, because e-governance tools will be used to enhance the power and control over information that policy élites possess over ordinary citizens;

- loss of control by politicians over the decision-making agenda in favour of control by a civil servants' or officers' 'infocracy' (Zuurmond, 1998) or an oligopoly of private contractors who supply e-governance systems or who supply the data management systems for services and transactions on which e-governance tools will draw (Margetts and Dunleavy, 1995); and/or
- loss of civil servants' capability to exercise constraint upon the populism of politicians;
- or alternatively, bureaucratisation, as decisions that can be made using algorithms are pushed down the hierarchy from politicians to salaried officials, because they cease to be regarded as policy matters at all (Bugler and Bretschneider, 1993);
- the supply of additional and expensive personal benefits for policy makers as they insist on access to those e-governance systems that benefit them, which add no great value to policy making itself (Nedovic-Budic and Godschalk, 1996; Berry *et al.*, 1998);
- the erosion of meaningful relationships among decision makers because of the loss of face-to-face contact (Wilson, 1999);
- even in the short term, the loss of quality in decision making because the focus on quantitatively measurable dimensions leads to neglect of less easily measured aspects of public programmes (Power, 1997);
- the loss of commitment to decisions made on the advice of computer based neural net models and expert systems, due to the reduced user control and understanding of the model, and the larger numbers of options generated (Landsbergen *et al.*, 1997); or
- in the long term, the erosion of humane values, because e-governance tools will bring about a culture of abstraction, rationalism and utilitarianism in policy making – Weber's 'iron cage of rationality' (Weber, 1958, 1976; Winner, 1977; van de Donk, 1998); the most extreme version of this view is that government technologies and rationalities, including e-governance tools, have become steadily more invasive through their extension of knowledge and information about populations, and their exercise has supported socialisation steadily to shape citizens into more docile subjects of authority (Burchell *et al.*, 1991; Barry *et al.*, 1996; Dean, 1999; Rose, 1999).

None of these claims is wholly compelling, and often depends on highly selective use of anecdote.

The bureaucratisation thesis, for example, is vulnerable to the powerful counter-example of monetary policy: the automatic correction mechanisms popular in the 1980s have been largely abandoned and the importance of judgment by policy makers once again recognised and given both policy emphasis and institutional expression in the remit given to central bankers.

Again, evidence for loss of meaning and of commitment is rather anecdotal, or confined to experiments, and does not necessarily support generalisation. Sometimes, norms, customs, rules, roles, practices that have worked well in face-to-face or telephone settings will be carried over successfully to the groupware context. This was the finding of a study of a network of British general practitioners in the National Health Service who began to exchange information about a range of practice management, prescribing, purchasing and health policy issues in which they were involved (Fox and Roberts, 1999). This finding seems to be explained by the particular institutional setting within which the online network was embedded. But it cannot be assumed in advance that this transfer will occur automatically. One study attempted to test for such changes by predicting what roles a group of people in an experiment would play in such a setting, based on how they behaved in face-to-face groupwork, and what roles they would want to delegate to the software, and then compared these predictions with roles observed. The main finding was that people played fewer roles than expected, and that at least some – but, to the surprise of the experiment designers, relatively few – participants described the software as playing the roles of recording the group memory and monitoring procedure, and a few saw the software as group facilitator and motivator (Zigurs and Kozar, 1994). It may be that, unlike the medical case, the institutional setting within which the experiment was embedded was simply too thin to provide existing norms that could be carried over. It is very difficult to avoid Hawthorne effects (people behaving differently, and often behaving as they think they are expected to do, just because attention is paid to them in the course of the study) in designing experiments of this kind, and even in designing studies in working organisations, and so it is hard to know what weight to put on such results.

So far, the move to reduce arbitrary administrative discretion by street level bureaucrats in public service decision making about individual entitlements such as cash benefits (stressed in the recent e-governance literature by Snellen, 1998), has not been the thin end of the wedge for the substitution of automated decision making in policy judgment, or the final coming of Weber's 'iron cage' of bureaucratic rationality (Weber, 1958, 1976). Despite the burgeoning investment by governments in the technologies listed above, there is little evidence of policy makers in finance ministries or land use planning departments following wholesale and mechanically the recommendations that are cranked out casually for them by junior staff from often still crude software models.

Noise, fragmentation and the erosion of reason

The third view is the most pessimistic. Here, the claim is that e-governance will actually erode rationality generally, both technical and political. It points to the pathologies of excessive demand for policy analysis thus delaying action – 'paralysis by analysis' – the bloating out of the policy advice industries among think tanks and consultancy firms, the problems of sheer information overload, the allegedly lesser ability of the public sector to manage information well than of private citizens or businesses, and the obsession with the already measured that distracts policy makers' attention away from tacit, implicit, qualitative, unstructured factors and toward formal, explicit, quantitatively measured, structured factors and information. Finally, and perhaps most crucially, this theory fears that, due to mechanical rule following as suggested by overly simple data interpretations, overly simple modelling, and by overly simple expert system flows from analysis to recommendation, the cultivation and the exercise of judgment in decision making will be crowded out. This view wholly rejects the cybernetic faith that information is control, and prefers the trope of information as noise. The main theoretical statement of this view in the literature on e-governance is given by Frissen (1999).

Much of the evidence for the noise theory comes not so much from studies of use of computers by policy makers, as from studies of accountability in public management. Successive studies of the actual use made of performance indicators – which are, of course, now typically collected, coded, submitted, analysed, collated, and now published electronically – have found that much of the information is not read, not used at all, or provides more complex pictures than those who hold public services to account really want or think that they need (Carter et al., 1993; Grønbjerg, 1993; Heinrich, 1999). However, this is too slender a base on which to erect a general theory.

Again, if one asks a different set of questions about coping with workload from those asked by many studies that find satisfaction with growing control, then by the power of suggestion by the phrasing of a question (e.g. using leading phrases that hint as 'information overload'), one typically gets data that can be used to support the noise theory of information. However, this too depends for such evidence as supports it, on answers to questions, where the answers given are highly sensitive to the ways in which the questions are framed.

Technology as used to reinforce institutional pressures

The theories in the fourth and final cluster of answers argue that there will be no very fundamental and independent impact of technology in and of itself on either the technical or political rationality of decision making. Rather, both continuities and changes in governance are driven politically and institutionally, and technologies are means, occasions and arenas for conflict over preservation ('conservative' social and political shaping of technology) or change ('radical' social and political shaping) of styles of governance are changed or preserved. (On social shaping of technology generally, see Mackenzie and Wacjman, 1985; Bijker and Law, 1994; Jasanoff *et al.*, 1994; Bijker, 1995.) Indeed, e-governance technologies are not wholly 'neutral'. Rather, design responds to social pressures in order to make social organisation embodied and implicit in the technologies (Akrich, 1992), but the degree of specificity and constraint imposed by technologies upon the social and political organisation of their users varies widely, both with technology and social organisation. Many technologies – the telephone for example – can be used to support almost any kind of organisation. In general, however, most technologies can be used to sustain a variety of ways of organising. This view is often described as using e-governance technologies as 'reinforcement' for what people, in particular institutional settings in the policy process, want to do anyway (Kraemer and Dutton, 1982).

For example, it could be argued that recent Dutch research on e-governance which shows its association with the promotion of horizontal co-ordination styles of policy making and public management associated with the adoption of e-governance tools generally, and modest experimentation in public consultation (Frissen, 1999), shows more conservative than radical, social and political shaping at work. While the rigidities of the traditional pillars may be almost dismantled by now, the Dutch tradition is perpetuated, of relying upon relatively autonomous networks of organisations in both implementation and rule making, which are not dominated by politicians. By contrast, the British case shows conservative social shaping of the opposite kind: technologies of e-governance are selected and used to perpetuate the British commitment to the dominance of the political executive, and second, to continue the role of elected politicians as story-tellers and leaders in public. Therefore, we find British politicians committed to seeing the development of digital television and to chat rooms – mass media for which politicians already possess the requisite communication skills – and to more leadership-centred technologies of e-governance for decision making. However, the prospect that most citizens might access digital services using mobile phones, which provide far fewer opportunities for the kinds of political communication that politicians need to engage in, rather than digital television, is one that many British politicians privately find disconcerting. All this reflects the prior political culture, at least at national level, that characterises the Westminster system.

On this view, then, the institutions, the symbols, the rituals in which e-governance technologies are embedded are what is crucial, not the technology itself, and they are what we should look to, in order to identify consequences. The possibilities of rationalisation, of an iron cage or of noise will only occur, if those are the effects of special but prevailing political forces and institutions.

Historical and ethnographic work on long run usage by policy makers of information technologies tends to provide more support for the institutional shaping theory (Nidumolu *et al.*, 1996), as does some survey work with a less conventional design (e.g. Overman and Loraine, 1994). Ethnographic research on the use of e-mail provides further good evidence that all projects for social shaping provoke counter projects. One piece of research shows that 'bandwidth' in use is as much a function of interpretation, implicit and tacit knowledge as of technical characteristics of the medium (Lee, 1994). Another recent major study on the use of electronic communication technologies for micro-political activity shows that even a supposedly 'simple' and 'low bandwidth' technology like e-mail does not greatly constrain its users. For, that study found, e-mail is used by rival groups, subaltern and superior, and opposed factions, to their own ends: that study shows that each exploits certain features of the technology – multiple addressing, recordability, processing and routing – to pursue their own goals (Romm, 1999). In this sense, e-mail merely reminds us of the longstanding finding in organisation studies that the flows of power, information and knowledge in organisations almost never follow the official organisation chart (Scott, 1992). E-mail does not deepen or exacerbate these informal networks of power and information, but merely provides them with new means to pursue their particular goals. This suggests that the hypothesis that because of e-mail, certain groups of policy makers will gain power at the expense of others, because they are gaining control (rationalisation at a price, perhaps?) – perhaps local chairs at the expense of leaders, or ministers at the expense of premiers, or civil servants at the expense of politicians generally – should be rejected. The technology is almost certainly not responsible. Rather, people use technologies as occasions, foils, symbols, tools, ritual objects, to develop and institutionalise the forms of politics and policy making to which they are already committed.

If there are changes in norms, customs, practices associated with moving to computer supported decision making among policy makers, that may at first sight provide evidence for the rationalisation at a price theory, we should be alive to the possibility that what is really taking place is radical social shaping. That is to say, the introduction of new technologies into groups of decision makers is also being used as the occasion and the opportunity for making other changes that superior decision makers want to make. Indeed, the conventional management literature argues that this is exactly what a good manager should do with technology, in order to avoid wasting investment on automating bad old ways of doing things (e.g. Hammer and Champy, 1995). There is an extensive body of work that shows that where superior decision makers use the opportunity to introduce more control, surveillance and accountability upon the work of lower decision makers, the technology will be experienced as introducing rationalisation at the price of discretion, autonomy and freedom of action (Shapiro *et al.*, 1991; Overman and Loraine, 1994; Hayes and Walsham, 2000). Academic researchers, being peculiarly sensitive to threats to their own intellectual freedom, can often tend to be more sympathetic to claims to autonomy in decision making by other groups than, for example, national politicians or national civil servants are, and this can lead them to represent political shaping of certain kinds – whether conservative or radical relative to institutional starting points chosen for the period of a study – as rationalisation at a price. However, many of the findings support the view that projects for radical social shaping of decision systems typically provoke counter-shaping projects of a more conservative or differently radical nature (and vice versa), because users are ingenious in using such systems for their own private benefits, subverting the intentions of superior decision makers who may be seen likewise as shaping the technology as much for their private ends as for organisational goals (Shapiro *et al.*, 1991).

The noise theory's reading of the findings about the non-use of indicators is not uncontested, and much of that evidence is better read as evidence for the institutional shaping account. For example, many recent commentators looking at audit and accountability mechanisms in the public sector see these as evidence for institutional shaping – namely, that rituals of accountability are being performed, symbols displayed, without necessarily achieving the substance of control (Power, 1997; Hogwood *et al.*, 1998). One important substantive effect, these accounts would suggest, of the development of e-governance tools principally to handle quantitative information has been a reallocation of that scarcest of resources in government, namely policy makers' attention (March and Olsen, 1976), to those problems that lend themselves to highly formal treatment. Had there been, for example, greater development and use in government of charting, creativity and soft systems methodology tools (Checkland, 1989; Checkland and Scholes, 1990; developing ideas from Vickers: see Checkland, 1994 and Vickers, 1995 [1965]; Ballantine and Cunningham, 1999), then attention to problems might be allocated differently, but that selection of technologies would only have taken place had the institutional pressures upon policy makers been rather different.

Precisely because the noise and institutional shaping theories differ in their basic conception of information from the qualified cybernetic consensus of the first two, it is possible for them to come to opposite conclusions about the impacts of new intellectual technologies on leadership and collective action from their conclusions about the impacts on decision making. For the very characteristics of being either noise and totem that can undermine the quality of decision making may in fact *increase* the capabilities of politicians to engage in leadership and collective action, for the cover of fog and the power of symbolism are powerful tools in the pursuit of these activities of governance.

Figure 5.1 summarises some of the key differences between the theories.

theory (specific view)	rationalisation	rationalisation at a price	loss of rationality	political and institutional shaping
information as effect on	control (positive)	control (negative)	noise	foil, arena, means
public management	integration	integration	fragmentation	depends on balance of forces
planning style	anticipation	anticipation	resilience	depends on balance of forces
shift of power arising	to policy makers	captured by special interests	undermining power	depends on balance of forces

Figure 5.1   Theories of e-governance at a glance

It will be argued throughout this book that the institutional shaping and 'reinforcement politics' view of e-governance is the correct one. That is to say, to understand the consequences of e-governance, we need to look at the *organisational* drivers of change in governance, to the service of which technologies are harnessed. We need to develop an organisational account of how information systems are used,[5] and we should reject overly simple information-driven accounts of how the organisation of the policy process will change. Unfortunately, informational theories of organisation are the fashionable myths of the present period (for a well-known example, see Tapscott, 1997b).

Unfortunately, this general finding about the social shaping of technologies does not get us very far. What we need to know is who shapes which technologies in which ways, for what reasons and on what basis. Developing a specific account of these processes is the principal work of Chapters 8 and 9.

There is, therefore, no general answer to the question, what will be the effect upon public policy of the wider use of e-governance tools. The 'price of reason' and 'noise' accounts reflect important fears about e-governance. In Chapters 8 and 9, something more will be said about just who fears these things, and why. However, the argument of this book in general is that e-governance software and hardware are the appropriate things upon which to pin either the hopes of those who would improve policy making or the fears of those who think it so fragile that technological change can only be for the worse. If we want to change the policy process, then we need to change the institutional settings in which e-governance technologies are embedded. In order to do that, we first need to understand them, not least in order to understand the scope that institutions of different kinds afford for change. The next chapter sets the scene for the presentation of such an understanding by considering three cases in more detail.

6
Conducting Policy Judgment through Information Technologies: Three Cases

Introduction

It is common to hear calls for more 'evidence-based' policy making: many cases of the use of information technologies in support of policy formulation are presented as 'improving' the linkage between the analysis of evidence and the design of policies or particular decisions. For example, early in their first term of office, Britain's new Labour government published a programmatic document setting out an aspiration upon ministers and their departments to make greater use of statistics and 'best available evidence' (Strategic Policy Making Team, 1999); only a little later, another programmatic document was published as government policy with the imprimatur of the Prime Minister calling for the greater use of modelling (Performance and Innovation Unit, 2000a).

Social scientists have consistently argued for many decades that while evidence of some kind always plays some sort of role in most policy decision making, rarely if ever can it be said that policy recommendations are accepted simply because they can be claimed to follow unambiguously from research or evaluation: what is counted as 'evidence' (in the broad meaning of information that is relevant to understanding a problem and deciding, however implicitly, upon action or inaction in relation to that problem) differs sharply between different groups in the policy process, and the manner of their selection and use of different pieces of information is highly contingent upon their institutional situation in the organisation of the policy process (6, 2002a; Weiss, 1979, 1998; Bulmer, 1987; Barker and Peters, 1992; Peters and Barker, 1993; Davies et al., 2000; Pawson, 2002).

In order to understand the way in which institutional features of policy-making bodies shape what counts as evidence in order to produce the styles of decision making that are viable within their institutional constraints, it is helpful to present some case studies which show some of the principal ways in which organisations make use of information technologies with which to manage, select, present and think about information that might be a candidate for the status of 'evidence'. These case studies will be used to illustrate the subsequent argument.

In this chapter, three case studies are presented, each representing a distinct style of organisation in its process of coming to judgment and of using information technologies to support that style. All are from the United States: each is based upon interviews conducted in November 2002. The staff in these authorities agreed to be interviewed under conditions of individual anonymity, which have been respected. These are two very different local authorities, one urban and one predominately suburban, and one federal department. The technologies used range from fairly straightforward spreadsheets and charting utilities through to large formal models, including geographical information systems and meeting management tools. In each case, the effect of the technology upon decision making is shown in relation to a function of the institutional character of the accountabilities being inculcated within the organisation, or in its relationship with other bodies in larger decision-making processes. Detailed analysis of the nature of these institutional styles and accountabilities is left for later chapters.

Baltimore city's CitiStat programme

Baltimore's CitiStat is the best known – indeed, celebrated – of the initiatives discussed in this book. It has attracted a good deal of press coverage in the United States, been adopted or at least adapted by many other cities; in part, this reflects the intensive promotion given to it by the political leadership of Baltimore; in some measure, it is also the effect of the promotion by national bodies within the Democratic Party such as the Progressive Policy Institute, and the recognition given to it by Harvard's Kennedy School innovation awards.

In essence, CitiStat is a system of decision making for improved performance management in city governments. At the heart of the system is a very particular kind of performance review meeting. It is of interest to the present argument because of the ways in which some very straightforward and commercially available software and hardware technologies are used to support this review meeting, and to add in a qualitative way to the distinct political and managerial experience that the review process represents for elected and career officials in the city government.

Martin O'Malley was elected Mayor of Baltimore in 1999. He arrived with a deep frustration about the quality of the public services provided by the city, and a suspicion of both the calibre and commitment of the managers and many of the staff that he inherited. Both in his campaign and subsequently as Mayor, he gave great emphasis to the alarming trends on crime in the city, as well as underperforming basic services such as highway maintenance, refuse collection and disposal, and rented housing management. On taking office, O'Malley wanted an initiative that would elicit rapid change and instil in the officials of city government his own sense of urgency and impatience. This fact more than any other has coloured the CitiStat programme, its design, culture and consequences.

The origins of the programme lie in O'Malley's conversations with one Jack Maple, now deceased, who had during the 1990s developed a similar programme for the New York city Police Department (NYPD), called CompStat. In essence, CompStat involved the detailed collation of performance statistics on a regular basis about every aspect of police activity and about the distribution of incidents to which the NYPD were expected to respond, their presentation in graphical form using fairly simple geographical information system (GIS) mapping tools, and the conduct of performance review meetings in which the geographical distribution and levels of activity were compared with the distribution and incidence of crimes and other events calling for police intervention. The idea was 'to put the cops on the dots', or match effort to reported incidents. CompStat was, then, a performance management and a resource allocation review system. It was a programme to which the then New York Mayor Giuliani attached great importance (Giuliani, 2002). Having been briefed about CompStat, O'Malley wanted to adopt it in Baltimore but also to extend it across the whole of the city government's services (for a managerial account of Compstat and its relationship with CitiStat, see O'Connell, 2001).

The CitiStat programme is the result. (The city provides information about the programme at http://www.ci.baltimore.md.us/news/citistat/index.html.) At the time of writing, there were some fifteen agencies within the purview of the CitiStat system, reporting either on a fortnightly or a monthly cycle.

Prior to each review meeting with a departmental chief, a highly structured process is followed. When an agency is brought into the programme, a template for reporting is settled. Many items are standard ones, set by the Mayor's CitiStat team: these include levels of staff sickness, disciplinary actions, accidents and untoward incidents, current staff headcount, some expenditure indicators and the like. Others are determined separately for each programme and the agency has some scope to negotiate the indicators selected, although of course the final say rests with the Mayor and his staff. Although templates vary in the proportions, there is a mix of input, activity and output and outcome indicators. The agency then supplies data according to this template, in the form of a spreadsheet file with raw data, about a week in advance of the meeting, to the CitiStat team in the Mayor's office. The CitiStat analysts then prepare a briefing report for the Mayor, using these data – and indeed any other information they may ask for in addition; the CitiStat analysts also have direct access from their own desktop computers to agency data and to data from the city's own customer call centre for reporting incidents and complaints. The briefing consists typically in between twelve to fifteen pages and may contain six or seven charts: the analysis is not mathematically advanced, being intended for the use of politicians and other decision makers who are not assumed to be familiar with complex statistics or algorithms. The report is not shown to the agency, which is assumed to be as capable as the CitiStat analysts are of working out what kinds of questions the Mayor and his team will be most interested in, and of performing the analyses that would best explore those questions. Reflecting the Mayor's own interests in differences in performance between districts in the city, a high priority is placed upon analyses using GIS-generated maps. The format of the report is standardised, so that if the Mayor does not have the time to read it before the meeting, he will be able quickly to turn to the relevant pages to find the information that interests him most.

In most cases, the available data only permit trends to be analysed back to 1999: however, this shortness of institutional memory also serves to reinforce the aspiration of the Mayor that the CitiStat process should represent a new beginning for the city government.

The analysts do more than produce statistics from agency numbers. They are encouraged to get out of their offices and check on the claims made by agencies. Equipped with digital cameras, they make unannounced visits to depots and local offices looking for bad practice or idling, drive around looking for potholes in roads that have not been treated. All this is used in evidence during the meetings in the CitiStat room.

During the meetings, a presentation is made, not by the agency but by the CitiStat analysts using the agency's data. The agency chief is then required to answer questions from the Mayor, deputy mayor and the CitiStat team. In the room are either heads or their representatives of the finance and human resources departments, the city solicitor, and of other central functions. Their presence is explained on the ground that agency chiefs may exculpate themselves by pointing to problems of finance, personnel or legal constraints that are beyond their powers to control, and these can either be ironed out in the meeting, if they are genuine problems, or else dismissed immediately if they are spurious. Meetings last about an hour and a half.

A special room has been designed, equipped and laid out in Baltimore City Hall for the CitiStat meetings. The cost of refurbishment and equipment, hardware, software is said to have been a relatively modest $20 000. Described not unfairly in some of the press reports as a 'star chamber', the room is laid out for confrontation. The ritual space for the meeting is organised along the two longest sides of a rectangle. A podium stands on one side, in front of a screen. The rostrum is occupied by the manager of the service under examination and required to justify her or his department's performance. Facing the podium is a series of chairs and desks for the questioners and representatives of other key departments. In the centre of this line is the Mayor's own position: to his left and right sit the deputy mayor and the director of the CitiStat programme. Behind them is a booth for the technicians who operate the presentation technology for the GIS system, the spreadsheets and other tables, PowerPoint and other software.

Mayor O'Malley believes firmly in the principle that individual area managers are individually responsible for performance in their area. Many GIS maps put onto the screen in the CitiStat room therefore show the name of the area manager for each district in the city for the agency in question, and these districts are shown colour coded by performance band for the category of performance being examined. More than this, the technicians are often asked to overlay the photographs of area managers onto the maps. The Mayor is said to call this 'putting a face on the problem'.

During the meeting, the Mayor or one of the CitiStat analysts may well check on claims made by agencies by telephoning an agency's call centre or local office posing as a local resident reporting an incident. The conversation is relayed simultaneously to the whole meeting.

The city has a contract with a company operating a geostationary satellite with the capability to provide satellite photographs in great detail of the whole of the city, and this facility too is used in the meetings. The CitiStat analysts tell the story with some glee of an occasion when one agency chief denied that their agency possessed a certain type of truck, which the Mayor believed he had seen on his own car journeys around the city. The Mayor was able to use a satellite photograph taken only a few minutes before the meeting to demonstrate that the trucks were indeed to be found on a site operated by the agency.

The aim, the analysts insist, is not to get agency chiefs to find solutions to the particular cases identified in these meetings but to elicit innovation from them in finding solutions to the systemic problems of which these examples are symptomatic: it is, they say, those managers which respond in this fashion who keep their jobs and thrive at O'Malley's Baltimore.

The confrontative character of the review meeting is made clear by the language used by the CitiStat team to describe it and the tools used in it. One said that the ability to relay live telephone conversations, to use digital photographs, to individualise responsibility on the GIS-generated maps, and to use satellite photographs meant that 'it gets really fun' and that for the agency chiefs there is 'not much opportunity to hide'.

The data too are presented with a certain confrontative authority. One CitiStat analyst told me, 'A map knows no loyalty. It just presents the facts.' The concern is here made palpable that the system should challenge any excessive loyalty that departmental chiefs might feel to their agencies.

The individualisation of responsibility is followed through into the incentive and sanction system in Baltimore, although there are some asymmetries between rewards and sanctions and between the agencies and the CitiStat team. Several agency chief officers are said to have lost their jobs in the early months of the programme, in part as a result of shortcomings that were revealed in the course of these meetings. High performers among individual managers are identified: the main visible rewards offered to them are free tickets to sports events, although it is thought that there are indirect effects upon the longer run promotion prospects of being identified in the CitiStat room as a high performer. They are, however, not given pay increments nor are there rewards for high-performing agencies as a whole, through their budget or otherwise: one Citistat analyst said, 'the punishment for doing well is that you're asked to do more'. Interestingly, though, the reports on the agencies are not individually signed but appear as the work of the CitiStat analysts collectively.

The Mayor now attends several meetings each week at least for the first part of a meeting, although in the first few months of the programme, he devoted much more time to them in order to impress agency managers with his seriousness of purpose about the programme. He always attends the meeting for the police department, because of his concern about community safety. There are very few weeks in which O'Malley does not attend at least four meetings for at least thirty minutes of each one: he receives briefings and minutes of all the meetings and is thought at least to skim read most of them.

One might be surprised that a mayor of a city as large as Baltimore can or does devote so much to reviewing operational matters. The CitiStat analysts reject the idea that the Mayor and his CitiStat team are 'micro-managing'. 'It's not that we are managing them and saying, "You should put Crew 4 over here" ', one said, 'No. It's that we're saying "Haven't you thought about that? Why is Crew 4 *not* over there?" ... We want to make sure they're on the job. It's holding them accountable that *makes* them manage.'

This form of individualised accountability under conditions of some suspicion is what defines the form of policy judgment which the CitiStat programme engenders. The role of technology is to provide a material basis for sustaining a style of decision making. The digital tools are, as O'Malley freely admits, much less important than what goes on during the meetings in the specially laid out room. More exactly, perhaps, the technology gains its particular meaning for decision making in Baltimore only from that use.

Fairfax county

Fairfax is one of the more populous counties in the state of Virginia with 370 000 households, located on the edge of Washington DC. Its jurisdiction includes some less advantaged areas but it also contains zones of high technology businesses, many of the employees of which live in the county. As a full service authority, it has responsibilities for health and social (or, in US parlance, human) services, education, transport, the administration of justice, policing, public safety, public works, planning and development, urban search and rescue. The county government's annual expenditure runs to more than $2.5 billions. Virginia operates a Dillon-rule system for county governance, which – by contrast with 'home rule' states such as neighbouring Maryland – provides that local authorities may engage only in those activities for which state law expressly provides: this constrains county policy making to some extent. The county's revenues come from a real estate tax, a personal property tax, a business licence tax and a sales tax; grants from the state support some health and human services, courts and police functions.

The elected members constitute a board of just ten 'supervisors'; nine are elected to represent districts, whilst the tenth is elected 'at large' as chair of the board. In addition to the county executive, four deputies are appointed subject to the approval of the board with executive management responsibilities for the main programme areas. Electronic government of every kind is given such a high priority, that one deputy executive, David Molchany, has responsibility exclusively for this work, working mainly with the Department of Information Technology. Fairfax has a reputation both within the United States and internationally for the sophistication and range of its online services, and Molchany himself is widely acknowledged as a leader in e-government.

In e-governance, however, it is for two particular capabilities that Fairfax is notable: the county government has made more imaginative and extensive use of geographical information systems (GIS) than most local governments, and its group decision support centre (GDSC) is an important contribution to its decision-making processes at many levels. Finally, its style of using digital information for performance management provides a revealing contrast with that of Baltimore City government, which was considered above.

Fairfax in decision-ready maps: geographical information systems

GIS constitutes a whole department in Fairfax county government. Awards are made to other departments which make the most imaginative use of GIS-based services, and in Fairfax, the uses are many. There is no general top-down strategy for the use of GIS, nor is there a system of rationing of GIS department resources between county departments or between problems. Rather, the strategy is deliberately an emergent one. All staff are encouraged to approach the GIS teams with projects and requests, and no employee needs special authorisation to do so. Throughout the county government, staff are encouraged to think geographically about problems and decisions.

For example, consider the case of the human services department as its managers prepare where to spend money to locate services, for example, to meet the needs of children at risk of violence, abuse, neglect and so on. Data on the geographical location of residences of such children can be shown on GIS-generated maps, by comparison with differences between administrative districts in levels of expenditure and/or numbers of staff or plant and facilities. Again using GIS, a major exercise was undertaken to redesign the routes used on each day of the week by county refuse collection crews, in order to improve efficiency, speed and responsiveness.

These might seem rather technocratic decisions. However, the location of the recently built southern county government centre was determined following an exercise carried out using GIS to provide information about the relative accessibility of various candidate sites for many of the services and professions that would work from it, and for various kind of groups of residents who might travel to it. Described as 'just so political' by one staff member in the GIS department. New routing for refuse collection could be decided at least in part (but only in part) like a mathematical optimisation exercise for a 'travelling salesman' problem; even here, of course, the decibel level of political demand from different groups of residents for early or late collections could not – and no doubt should not – be ignored. In the case of the southern government centre, the selection of scenarios was highly constrained by the availability and affordability of suitable real estate, the complexity of public perceptions in different areas sometimes demanding additional services and at other times resenting the presence of facilities that might be used by poor people and the concerns and lobbying effort of individual supervisors. Such local concerns were even more significant when GIS systems were used to model the implications of various possible relocations for neighbourhood fire services. The constraints with which the planners sought to work included the target number of minutes response time, modulated between peak and off-peak traffic times, and assumptions about likely speed, the number of houses served by each station: of course, each of these assumptions was the subject of extended debate and political interest, and many trial runs were required before acceptable settlements could be struck.

The administration of welfare services also presents opportunities for the use of GIS to support or, more exactly, represent each stage in what are in fact highly political decisions. Food stamps, for example, used to be distributed at the county government centres, creating long queues: the question arose whether they could be made available through commercial grocery stores instead. But which ones? The GIS modellers coded every grocery store and its location, and every food stamp recipient, and made some guesstimates about how many people might use each store, recognising that people would not necessarily use their nearest store but might well collect their stamps at other stores while travelling around the county. The possibility of queues of claimants reappearing had to be avoided, especially in areas where local residents might react badly to this. Not every grocery store could be assumed to be willing to participate in the scheme. Following the introduction of a general welfare-to-work programme following the 1996 welfare reform law, the GIS modellers had to work out how and where to locate new investment in childcare facilities for parents moving back into work or into work for the first time, taking account of their location and clustering and of current and planned bus routes, the implications for congestion on certain routes into and out of particular areas at certain times of the day, and possible concerns by better off groups at the inflows of stigmatised groups into areas where they had not previously been seen in significant numbers.

The wired horseshoe: group decision support

The Group Decision Support Centre (GSDC) in one of the Fairfax County government centres is modestly staffed, with one coordinator and just two others on a full time basis: the coordinator and one of the others are qualified facilitators of meetings that use the centre, while the third is an administrator. The desks in the room are arranged in a horseshoe, and can accommodate up to eighteen people in a single meeting. Each desk includes a personal computer connected to a local network and equipped with a range of meeting management tools. The facility can be booked by almost any team of people charged with making decisions.

It is in use most days by some team of managers or other. Like the GIS programme, there is no county-wide strategy – neither of mandating units to use nor of rationing the resource, beyond the simple 'first come first served' allocation of days.

Most meetings last a day or less, but where justified, a session might last longer, and some really major problems might involve a series of sessions each separated by a week or so.

The software allows people to debate an issue with everyone contributing simultaneously and in text. One of the advantages is that everyone can 'speak at once'. Another is that contributions can be anonymous, so enabling greater candour – although in many meetings it may well become obvious later on just who has made which particular arguments. The software also enables complex sequences of voting on variously hierarchically related and alternative branches of options.

Common uses include internal reorganisations, purchasing designs and some strategic management sessions. The board of supervisors frequently appoints special commissions to analyse policy problems and to prepare recommendations: these bodies often make use of the centre in the later stages of their work to conduct the arguments that lead to consensus, or at least majority support for their recommendations.

A major exercise which involved repeated sessions in the centre by a group of senior managers was to make a decision on which customer relations management (CRM) system to purchase to underpin the work of the board of supervisors. This presented particular difficulties of lay-out and meeting management because it was necessary to conduct some of the meetings with potential suppliers in the room, while the decision makers had to ensure that what they wrote about these suppliers could not be overseen.

While in theory the board of supervisors could use the centre, they have not done so and the state's and the county's own 'sunshine' laws under which they operate would make it difficult for them to do so. All their meetings must be conducted in public: indeed, more than two of them may not meet nor may they hold conference telephone calls, without the meeting being publicly advertised, minuted and the residents of the county permitted to attend or listen. Not only is the GDSC space too small for the public to sit in on meetings, it is designed to permit decision makers to make some contributions anonymously and so to speak with greater candour. It is precisely these features that are effectively prohibited to the board by the sunshine laws. However, there is no reason why other elected officials elsewhere who are not subject to such rules could not make good use of facilities of this kind.

The coloured map inside the horseshoe: bringing GIS and GDSC together

The close imbrication of technologies of e-governance with close hand-to-hand politics is perhaps best exemplified in the way in which Fairfax county made its decisions on redrawing the boundaries of the electoral districts in 2001, following the publication of the census data. Whereas the United Kingdom and a number of other countries have taken the decisions about where to draw the lines on the map between the constituencies for elected representatives out of the hands of politicians and given the function to appointed commissions of politically neutral persons, in the United States, these decisions remain ones for political bodies, subject to challenge in the courts. The redistricting process involves the development by county boards of plans which are submitted to federal government for approval, which must comply with various regulations and standards, which prescribe equalisation of numbers of electors in each district according to census calculations. In many states of the United States, of course, redistricting is a peculiarly highly charged political issue, because of the centrality of racial and ethnic divisions in the general politics of the country, overlaid on other partisan divisions. Between the 1990 and the 2000 censuses, the proportion of Fairfax county's significantly increased population which was scored by census analysts as 'Asian / Pacific islander' and 'Hispanic' grew by much more than the African American proportion did. Overall, those not scored as 'white' grew from about a quarter of the population in 1990 to about a third in 2000 (see http://www.fairfaxcounty.gov/comm/ demogrph/othergrph4.htm). It is not an area of particular racial tension, nor does it include areas that most people would consider to be among America's 'ghettos' of seriously disadvantaged minority groups, but the national profile of issues of ethnicity ensures that local politicians have to tread very carefully in redistricting. Moreover, the legal requirement for roughly equal numbers of electors per district (more exactly, districts should be defined to include within 5 per cent either side of 100 000 voters per district) does – as in all democracies – come into conflict with other imperatives that boundaries should reflect, felt communities, and quite apart from issues of ethnicity, this gives redistricting decisions a certain piquancy.

However, just because politicians are charged with the responsibilities of making these decisions, it does not follow that American citizens are any more content than electors elsewhere with the thought that their elected representatives might arrange the boundaries to suit their parties' interests, or even their own. To satisfy public concerns, the Fairfax board decided that for the 2001 redistricting exercise, they would, as they or their predecessors had often done before, create a special commission to produce the first set of options and even recommendations, and they would require that commission to consult the voters (other counties' boards simply asked their staff planners to prepare the options). What was different about 2001 in Fairfax was the availability of both GIS and GDSC technologies to support the process.

Having conducted its rounds of face to face meetings and received written submissions from a variety of groups and individuals, the citizens' commission conducted a series of meetings in the Fairfax GDSC to prepare and debate the options. They were able to use the groupware to debate issues anonymously before returning to identified exchange, and to simplify the voting process. However, the major innovation was the introduction of the GIS system into the GDSC room on a huge screen in front of the horseshoe. Instead of having to ask county staff to go away and prepare printed maps of options for consideration at the next meeting, the GIS team were able to create maps on the screen of each option, on the fly. Naturally, this had the effect of speeding up the process, which was complete within a month of regular meetings rather than the previous exercise which would be completed in three or even four months. However, it also engendered a qualitatively different experience. The commissioners felt much greater confidence and sense of their own independence because they were able to see their ideas represented immediately.

The ability to offer this kind of GIS support raises afresh but in much more detail and with greater immediacy, many of the traditional procedural questions about what information should be presented to bodies charged with this kind of decision making. Should commissioners, for example, be able to see information about ethnic composition of possible districts only at the level of the possible district they are considering as a whole? Or at street level? Or at block level, since many debates about how far to move a boundary would be about moving a single block between districts? What about information about the numbers of elderly people? Or about socio-economic class? After all, each of these would give the astute commissioners some indication of the likely implications for partisan dominance of potential districts of each of the boundary changes they might make. How thick should the veil of ignorance be under which decisions are made? The regulations and standards provide some guidance, but they cannot of course be complete, since they are not specific about exactly what census information, and what other locally collected information available may and may not be taken into account.

In the end, the agreement was reached that information would not be given down to block level, and that age, and ethnicity indicators would not be used on the GIS maps. Clearly, one could debate the merits of the Fairfax decision endlessly. In favour of it, one might argue that these are precisely the kinds of ascribed categories that ought not, on some views of impartiality, to weigh with decision makers: privacy considerations can be used to support a refusal to use block-level information since a knowledgeable individual could readily re-identify such information. Against it, it could be said that in the absence of precise statistical information, the decision makers were in danger of relying on their hunches and presumptions or their experience of canvassing in particular streets for their understanding of the compositions of particular streets and areas, and this might well be error-prone and lead to worse decisions. After all, it is highly unlikely that the supervisors, when they came to consider the commissioner's options and recommendations, did not take account of this kind of information, however informal its character and sources. The merits of the particular decision the Fairfax county government and the citizens' commissioners came to in 2001 are not at issue here. What does matter is that the presence of the GIS system and the ability to generate maps with these kinds of information made it necessary to come to decisions about these things very much more quickly than in the previous redistricting exercises, and in conditions of information resources and capability that enabled these things to be presented on the fly and in the course of a process that was thereby much compressed in time by comparison with previous decades.

Perhaps someone might wonder whether the use of the GIS software to generate the maps and of the group decision support system for deliberation within the commission did not give a spurious air of technocratic authority to a decision that is, and should be seen as purely political in character. Even had the GIS technicians sought to represent the process in this way – and I have no evidence that they ever did and in the interviews I conducted they were careful to stress the political character of the whole procedure, it seems to me to be unlikely that anyone in Fairfax would have been taken in by such a suggestion. The first question that most Americans ask about any redistricting proposal, however produced, is certainly not 'what algorithm was used?' but the simply *'cui bono?'* or 'whose interest is being served?'

Performance management

Like Baltimore, Fairfax has a performance management system, which involves review meetings in which GIS is used. In every other respect, however, the two authorities could hardly represent a greater contrast.

No special room is used for the purpose in Fairfax, let alone a specially designed confrontation space. Indeed, in the main government centres in Fairfax – all of which are rather recently built, by contrast with the Progressive era city hall in Baltimore – physical space is highly flexible, generally open-plan, using moveable partitions rather than fixed walls wherever structural considerations permit this, and organised in ways that are designed to promote team working.

Whereas in Baltimore, performance is measured from the beginning of Mayor O'Malley's period in office, in Fairfax, data go back much further reflecting the much greater stability in both political and management personnel and styles. The performance review meetings are held much less frequently, and generally on a quarterly basis. They are chaired, not by the chair of the board of supervisors but by the county executive. At Fairfax, there is much less individualisation of responsibility, and much greater emphasis on team working within departments.

The county executive stopped one meeting quite early in the series with one department, declaring that the template of performance measures being used was clearly inappropriate: he asked for a new template to be produced in time for another meeting three months later. It is hard to imagine Mayor O'Malley readily taking that approach in Baltimore. More generally, the system of performance management is not seen as the principal instrument for bringing about service improvement in Fairfax, but as part of the whole suite of initiatives in which the county government is engaged.

Perhaps services are objectively much worse in Baltimore than in Fairfax; or perhaps the fact of recent political change represented by O'Malley's arrival and determination to make a difference quickly and his own perception of the calibre of services is a more powerful explanation of the difference. Or again, perhaps someone might think that at Fairfax, senior managers and board members are not sufficiently tough with agency chiefs. Or perhaps Baltimore city government has a generally more individualising and confrontative political culture between elected and salaried officials. Whatever the explanation, and irrespective of which approach makes most sense either generally or in its own context, the differences are very marked.

Policy judgment in Fairfax exhibits therefore the style of much greater team working. The ritual space of the horseshoe in the GDSC stands in the sharpest possible contrast to the layout of the CitiStat room, with its air of a criminal courtroom. The frequent use that the Fairfax supervisors make of special commissions to tackle specific problems, and their commitment to deliberative processes stands in marked contrast to the way in which the citizens of Baltimore contribute to the ongoing rather than episodic CitiStat process, which is on a strictly individual basis through their telephoned requests for services and complaints.

The US Federal Department of Energy

At one level, energy policy is global: the oil industry, for example, thinks, plans, invests and organises on a planetary scale. On the other hand, it is highly nationalised: the tax and subsidy treatment of petroleum, renewable sources and nuclear-generated electricity systems vary hugely between countries. Again, it is very local, as the unending series of planning disputes over the siting of wind farms show.

At each of these levels, however, several things are constant. First, it is essential to think over twenty-year time horizons. For the capital investments involved in almost every current kind of mass energy production and distribution systems (with some exceptions, which today play a limited role, such as single building solar panels) are so huge that unless one can be reasonably confident of there being a return over that kind of period, few would hazard their money – and that includes the representatives of the taxpayers. Beyond 20 or 25 years, it is very difficult to attach any acceptable levels of confidence to any numbers that could be dignified with the title of 'estimates' for demand or costs or behaviour or technology availability. Another feature is common to energy policy problems at every geographical level of aggregation: they are inextricably imbricated with environmental issues. Fossil fuels are thought to be one of the major sources – if not *the* major source – of gas emissions, which, most people now accept, are likely to be an important causal factor in global warming; at the local level, vehicle emissions of particulates are the subject of conflicts over traffic calming measures, the siting of schools and the like; at every level, people debate the merits and demerits of civil nuclear power plants in terms of the risks of release of radioactive material produced as a by-product of the operation of reactors, just as no one now discusses the merits of hydroelectric schemes without spending much of their time on the environmental impacts. Moreover, energy policy is highly professionalised. Engineers with many different subspecialisms have a privileged role in debates about energy policy, because of their near-exclusive command of knowledge about the rapidly changing technological and design solutions to many of the challenges of energy efficiency and the development of alternative solutions, in respect of technologies, scale, for generation, distribution, use and by-product processing. In recent years, new professions have gained authority in the field, including environmental lawyers, specialist energy and environmental economists, and even social scientists specialising in measuring consumer attitudes. Finally, at every level, energy policy is racked with uncertainties – about the prospects for demand especially for new technologies, about the responsiveness of consumers and suppliers to changes in tax and regulatory regimes, about the impact of shocks such as wars in oil producing regions, about environmental impacts.

These features – large capital investments, moderately long time horizons, complex interaction effects with environments, high degree of professional dominance of the field but growing numbers of partially competing professions, and multiple uncertainties – go a long way to explaining both the willingness of policy makers in the field to turn to modelling or simulation to help them understand how industries might respond to different combinations of policy instruments, and the emergence of capabilities to supply that modelling. A model, in this context, consists in a series – perhaps running into thousands or even tens of thousands – of equations representing ratios between variables, by which to describe how a system responds to changes. Represented in software, a model provides a policy analyst with a tool they can use in order to try to understand how the system being described might respond to policy changes their governing authority might make. Of course, that understanding is qualified in various ways, according to the degree of confidence that the policy makers have that the model is sufficiently detailed, accurate as far as it goes, the variables well specified, and so on, for it to yield a sensible answer to the type of question being put to it, and according to the calibre of the data being fed into it.

The use that the US federal Department of Energy makes of models of several kinds provides good examples of the vastly greater importance of political, than of purely technical features, in the processes by which modelling supports policy work. Moreover, just as the individualised and mayorally dominated politics of Baltimore city government shape the meaning of Mayor O'Malley's CitiStat programme, and just as the more collegial and relaxed culture of Fairfax county shapes its ability to use GIS and group decision support, so the micro-political organisation of the Department and its relations with other federal departments, and with the White House and other central agencies of the administration shapes what modelling means, and what policy staff will use it for, and what consequences its use may have.

Within the department, but created by separate legislation with a duty to exercise its functions independently is the Energy Information Administration (EIA). The EIA is custodian of some, but by no means all, of the modelling capability available to the department. Indeed, in those years where there has been great demand for modelling work, its services may have constituted a modest proportion of the total modelling effort commissioned. One of its largest modelling engines is the National Energy Modelling System (NEMS). This consists of many hundreds of equations representing supply, demand, environmental impacts, some technologies and the like.

Occasionally, a deputy secretary is appointed with greater interest in modelling and analysis, and who can secure permission from the President and the rest of the administration for a major strategic initiative. This was the case in George Bush senior's administration, when a huge modelling effort was undertaken for the National Energy Strategy, involving a much greater in-house capability than now exists in the department. Since then, in-house capability has declined significantly.

Many of the senior staff in the Department are long serving, and know each other well from many years of collaboration: many have been with the department since its establishment in roughly its present form in the 1970s. This in turn creates a large body of collective memory, much of which remains implicit but which nonetheless structures policy debate: few issues have no analogy within the working memory of the senior managerial policy, research and technical staff. This forms a core of – depending on just whom one counts – a hundred people at the headquarters on the Mall or closely linked with HQ.

However, senior managers and policy staff are not completely secure in their posts, even as career civil servants. Following a change of administration in the White House and the arrival of a new Secretary and other political appointees, an ambitious junior might hint to politicians that their erstwhile boss may have been a little too assiduous in working for the previous administration in the hope that the person might be moved, making a vacancy. This does of course erode some of the trust that otherwise binds the senior cadre together. In the last year before a Presidential election which is expected to produce change of regime, some departmental staff report signs of what might be called 'defensive policy work', or work that is much more carefully ideologically distant, or even apparently seeking equidistance from the commitments of the parties.

A responsible chart: the political organisation of statistics

What can be presented to the political appointees brought into the department by each incoming administration closely shapes what modelling gets done and what use is made of the results – both for high constitutional reasons of democratic accountability and for such simple practical imperatives that no one wants to waste work. The Secretary for Energy and his or her Deputy Secretaries are busy, have short attention spans, are not necessarily trained in very advanced mathematics, and want information to fight very particular battles within the administration or in Congress or the media. The kinds of questions they ask may not necessarily be the questions that, for example, energy researchers in academic institutions or in huge oil companies or in local groups campaigning against particular facilities might ask, but those questions will define what analysis and modelling the career civil service does. On the other hand, those career civil servants feel keenly their duties to provide honest and independent advice, and to provide multiple options wherever they can, even when they know that the political appointees have no interest in considering most of them.

How to strike some kind of settlement between the competing pressures, in the selection of modelling to be commissioned or reviewed and in the selection of information to be presented? The key terms by which the career civil servants working as professional analysts describe the standards that their analysis should meet in order to fulfil some kind of settlement are 'robust' and 'responsible' and 'useable'.

Much of the analysis done today is in fact synthesis of results from a wide variety of modelling studies conducted by the EIA, by other governmental research bodes such as the National Laboratories, by independent commercial research companies, by the energy supply companies, by academic research units. A common type of synthesis chart used by the department – and which is also used, for example, by the Bank of England, its regular syntheses of economic forecasts for the British economy – is to present a single data series on a historical trend in the form of a line representing a change in, say, energy prices, or greenhouse gas emissions, or demand fluctuations or even elasticities of consumer or supplier response to marginal tax changes leading to the point of the last date for which actual data are available and then for the right hand side of the chart, to present a trumpet of lines branching from the last date for actual data, representing the spread of projections, forecasts and scenarios in the studies reviewed. Because this presents a full range of possibilities taken seriously by the main modelling groups to which status is attached within the Department, it is described as 'a responsible chart'.

In selecting which of these lines to which to give particular weight, analysts take account of two factors, which might be summarised as 'authority' and 'consensus'. Where there is an independent, respected and scientific body in the field which itself conducts an exercise to review all the major studies in the field, and which tries (for example) to select between the projections on the basis of some procedure for attaching probabilities to their various assumptions, then its preferred projection would be given greater weight. In the field of impacts of energy use upon greenhouse gas emissions, the work of the Intergovernmental Panel on Climate Change (IPCC) would be privileged in this way. Analysts speak carefully of 'relying' on IPCC projections, but 'taking note' of others: 'taking note' means what it says – they are neither ignored nor privileged; they are presented alongside the authoritative projections, but their second tier status is clear. The second factor is the simple one of looking for where the majority of the forecasts cluster. Clearly, where the views of the responsible body and the majority coincide, as they often will, analysts will feel even greater confidence in advising the political appointees that they should pay most attention to this zone in the trumpet. Where they diverge, greater caution and heavier qualifications are called for. Where authority or consensus clearly indicate a projection or a narrow zone of projections to converge around, this is deemed to be 'a robust projection'.

However, in producing such a chart, in general, the departmental analysts do not include error bars or standard deviations, even if such statistics are available, and they cite two reasons for this. The first is that they assume that the political appointees will know that these uncertainties will be large in any case, and second, they have no reason to think that many political appointees would understand or recognise them. Many matters of deep uncertainty simply do not make it into a briefing: if a variable is one about which very little is really known, analysts would question whether it is appropriate to include it in a briefing at all. As uncertainty rises, the onus of argument in the department shifts sharply on to those who would consider that it is still vital to brief politicians about it. The reluctance of politicians to deal with more technical uncertainty than they absolutely have to defines the scope and role of analysis.

Moreover, in any briefing, they would rarely offer more than six charts, and never more than two lines on a chart, reflecting their experience of what political appointees can understand and are prepared to read. A few politicians are willing to accept briefings that use multiple scenarios. When Mr Gore was Vice-President in the Clinton administration, for example, he was prepared to use this approach, but he was an exception. A small number of scenarios represent the very outer limit of the 'useable'.

It is worth pausing to reflect on the contrast between the vocabulary of responsibility, robustness and usefulness at the department with the breezy assertion by the CitiStat analyst at Baltimore of a GIS-generated map that 'A map knows no loyalty. It just presents the facts': it is hard to imagine an Energy policy staffer making such a statement about one of their charts.

Perhaps to many readers, this reliance on authority, a tradition of previous analysis using a recognised method, and consensus will seem to be only common sense, 'rational' or even the only thinkable options. To other readers, perhaps, it may seem surprising, even flawed. For the benefit of those who cannot think of any other reasonable course to pursue, it is worth pointing out some of the potential limitations. Life is, after all, full of surprises and the conventional wisdoms often turn out to be wrong: shocks, unforeseen consequences, the insistence of consumers upon using technologies for their own goals rather than those expected by the designers, processes of drift and self-defeating processes, will all undermine the efficacy of reliance on the mainstream disciplines and the average view. Not infrequently in technology forecasting, those who have adopted these methods have been found wanting. For the benefit of the second group, it is worth pointing out that minority views, however wrongly discriminated against, have no greater guarantee of being right than have the sheep who simply follow their leader.

But this approach generating policy judgment through reliance on authority and consensus makes perfect sense for an organisation such as a federal department. The department needs to be able to defend the advice that it gives to political appointees as reflecting the broad swathe of moderate opinion among those who use the intellectual disciplines – engineering, economics, climate sciences of various kinds – by which such defence can credibly be mounted. That rootedness in the intellectual discipline is the hallmark of what makes a department different from, say, a think tank associated with one of the political parties or a group of journalists. On the other hand, they cannot offer all the hedges and qualifications that quantitative scientists might in the form of confidence intervals and standard deviations. They have a political client, and that client's needs must be respected and met.

In the department, then, charts are known not 'just' to present 'the facts'. The problem of uncertain and competing loyalties of charts that might be presented is settled by the compromise of offering summary charts that present all the projections, but weighting them by authority and consensus. A 'responsible' chart has multiple loyalties but reconciles them by covering the analyst in one respect – all the projections of 'serious' forecasters are there – but not in another – error bars are not presented, and no more than six charts each with no more than two lines are used. These compromises reflect the micro-political organisation of the department itself – professional, but accountable to politicians; a major player in the energy system, but attempting to secure its position as a major player by the consent of all those others with something to contribute and not just by brute regulatory power; structured around the duality of the professional team and the more individualistic political appointees, and charged with ensuring that the institution of the professional team can be defended, and that the political appointee can defend their particular case.

Battles of the models: the loyalties of simulations

Models alone have never settled any policy question in the department, and least of all in inter-departmental conflicts. For any set of results from a simulation run on a modelling engine is sensitive – to varying degrees that can be measured statistically – to the particular assumptions fed in. One departmental staffer put it, with appropriate amusement but without any trace of cynicism, 'That's the game that gets played around here. Anyone can make any model say anything and they do.' However, this is not an adequate description, for it would be wrong to think that modelling is irrelevant or merely expensive intellectual decoration upon brute arm-twisting. Departmental staff talk of modelling, and even 'battles of the models' as 'refining the debate'. 'Refining' here is more reminiscent of the refinery – that is, of fractionation of crude oil into distinct parts – than it is of refinement, in the sense of simple *politesse*.

Inter-departmental conflicts are the most testing environments for the politics of modelling. Typically, these erupt following an indication from the White House that the President would like to make an announcement or launch an initiative on a particular subject. We can consider the case of something that has significant environmental repercussions, whether for climate change generally, or perhaps a particular technology that might or might not prove more efficient to emit lower levels of particulates or have a smaller demand for sensitive land than another available technology, or than the ones presently in use. With a problem structured in this way, the ground is ripe for conflict between the Department of Energy and the Environmental Protection Agency (EPA). These conflicts are fought out in inter-departmental working parties and committees, in part through the use of modelling exercises.

In a battle of the models, a first stage is for each side to run their own preferred assumptions on their own models. Sometimes, both departments will use the same model: EPA sometimes use NEMS even in conducting a battle with the Department of Energy, if it is the most appropriately specified model. After all, in a battle for intellectual credibility, it makes sense to use the most appropriate engine, even if it is 'owned' by the opponents: in fact, since NEMS is operated by the EIA, and since the EIA is legally bound to be impartial, any departmental ownership by DoE is much attenuated. Next, each side then tries to second guess the kinds of assumptions that their opponent department would want to make, and then runs those assumptions in order to see just how sensitive the results are to each of the particular assumptions. Typically, each side will only see those modelling runs of the other department that they are voluntarily shown in the course of the inter-departmental working parties and committees: the rules of the battles do not permit espionage or quiet pressure on EIA technicians to reveal what work has been done for the other side.

This focus on assumptions and the sensitivity of results to particular assumptions is the first stage in 'refining the debate'. For it serves to crystallise and define the particular areas in which each side must develop the most sophisticated and credible case for its assumptions. That case is not generally derived from further modelling, although sometimes that might help. Rather, assumptions are defended discursively and qualitatively, using the usual armamentarium of policy analysis, ranging from desk research, review of literature, use of other statistics, appeal to historical analogy, cross-national comparisons, construction of plausible causal accounts, explication of likely risks and less likely shocks, design of additional policy instruments by which untoward behaviour by crucial variables might be prevented, and – when these things prove insufficient, appeal to basic values, selective quotation from existing policy commitments, use of public opinion data, and so on.

Battles of the models are deemed to have been settled in a limited number of ways. As the date approaches by which the President wants to make an announcement, pressure mounts for a resolution. Making a credible case for a set of assumptions means persuading key politicians or their appointees, either in the White House, or in other departments that may be important allies (Transportation and Agriculture are the other main interlocutors in most of Energy's battles of the models). In the last analysis, the credibility of the case for the assumptions will rest on what the politicians are prepared to try to sell in Congress, especially (but not exclusively, given the character of US bipartisanship and loose party discipline) to their own party and in the media.

International negotiations are not processes in which models or results generated by using them play any significant role. For in international conflicts, the discipline is entirely absent of each party being a component of the same organisation (e.g. a national government) and having to defend their position using arguments that are grounded in a common overarching policy stance and in certain basic policy sciences. Thus, were a negotiator to try presenting results from modelling in support of a position, the response they could expect would be a blunt dismissal of their relevance, truth or integrity.

Within the government, modellers in different agencies may have conflicting loyalties (e.g. a growing energy market or a shrinking environmental footprint) but they share certain common assumptions and approaches to their style of analysis and argument.

Taken together, then, the DoE exhibits a distinct style of policy judgment using information and information systems. In battles of the models, there is confrontation, but the ritual order of confrontation is very different from the individual cross-examination of managers using data taken to be self-explanatory, that is the hallmark of the Baltimore approach. There is collegiality as at Fairfax, but the political appointees and the instabilities in the post of the senior career civil servants do introduce a degree of individualistic organisation. The predominant style of policy judgment is that of professional and analytic inference from assumptions, using models and careful synthesis of evidence, following the main canons of policy analysis, subject to limitations accepted as deriving from the interests, patience and abilities of those political staff. Of course, all this takes place within the general parameters of the commitments of the department to the interests of security of energy supply, to avoiding causing huge upheavals to important US companies and to maintaining an acceptable fiscal stance. This contrasts – and occasionally conflicts – with some of the basic commitments to greener production and consumption which are the hallmark of the EPA, but the EPA too shares the basic analytic approach. By contrast with Fairfax, rank and status matter much more in the DoE. The status to 'work interagency' – that is, to be involved in negotiations on inter-departmental committees and working groups – is prized greatly, and is much more likely to be held by policy staff than by researchers or modellers. This subaltern status structures the role of modelling and the nature of reliance on information from models. With rising status comes growing insecurity around the time of change of administration, and this leads to more defensive applications of this style.

E-governance technologies, judgment and political organisation

This chapter has presented three very different organisations, each using very different technologies to support their policy making. In each case, a distinct style of policy judgment and political judgment can be discerned, which the technologies in question are used to sustain.

A technology such as GIS mapping can be used to support quite different styles of judgment. In Baltimore, it is used in order to conduct an adversarial process, in which responsibility is individualised, and in which decision making is focused on the operational level. By contrast, at Fairfax County, GIS is used very widely in almost every department, but in ways that uphold a more collegial style of decision making, in which there is much greater delegation of oversight roles from the most senior elected and appointed officials.

This is not to say that technologies in general are in any simple way 'neutral'. However, some technologies are more open to uses by contrasting forms of political organisation than are others. The general view that politics are so embedded in technologies that having once adopted a technology, people must then follow its implicit political prescription is surely vastly overstated (e.g. a view to which at various times Langdon Winner has veered [1980]; some interpretations in the science and technology studies literature of Bijker's 1987 concept of a technological frame also run close to this view). GIS appears to be one that is much more flexible than some others one could examine (possessing nuclear weapons, for example, commits a state much more irrevocably to a limited menu of ways of conceiving choices in foreign and security policy). Forms of political and administrative organisation are certainly not neutral between the uses of technologies, or even between technologies themselves. Given the organisational characteristics of a central government department such as DoE, and given the character of the field of energy policy, reaching for modelling as a strategy with which to conduct policy judgment is entirely understandable. Modelling did not deliver and could not ever have realistically been expected to have delivered the single indisputably correct energy policy. What it did for DoE and for its interlocutors in the energy policy process was, as officials put it, 'refine the debate' – precisely the outcome of things that a body organised by administrative hierarchy, profession and accountability to elected politicians would be expected to hope for.

Certainly, having adopted a technology then constrains future choices and styles of judgment to some degree. The Department of Energy, having found itself a user of modelling, then found that it conducted its political conflicts with other agencies through 'battles of the models'. Having adopted the reporting templates of the CitiStat, Baltimore city government is finding that there is a steady growth over time in the numbers of indicators being agreed to be added to these templates. In time, we can expect that the micro-politics between the centre and the departments will be organised in significant part around negotiations over the content of templates. In Fairfax, the commitment to a GIS department and to encouraging departments to make use of it has no doubt changed the distribution of attention to aspects of policy problems. The capability to examine geographical distributions of problems and of resources contributes to the institutionalisation of a style of policy judgment in which questions of territorial equity and efficiency are peculiarly privileged. Over time, the skills of asking policy questions in geographical terms has been learned and refined, and indeed disseminated around the county government. In Fairfax, someone who wanted to argue that too much attention was being given to these aspects of decision making would find it hard to gain a hearing.

In the chapters that follow, an account will be developed in much more detail of how technologies are used in different locations and styles of political organisation, to sustain distinct practices of policy and political judgment. The argument will be illustrated by reference to these cases.

7
Policy Making: the Work of Governance

This chapter is concerned with the question, what is it that policy makers do and need to do, that e-governance tools ought to be supporting? It will build on some of the insights that can be taken from the three detailed case studies in Chapter 11. The first part of the chapter argues that what policy makers do, first and foremost, is to try to make sense of a bewildering and uncertain environment. They need, therefore, not so much more information, as the right kind of information that helps them commit themselves to a course of action, and to reduce the range of available interpretations to manageable proportions.

In the second half of the chapter, this leads to a detailed analysis of what is understood by political and policy judgment. Five key elements of political judgment are distinguished, by reference to the orientations to time that define the basic commitments of exercising genuinely political judgment. This prepares the way for Chapter 8, which explores in detail the institutional constraints upon different kinds of policy makers in carrying out their roles in the policy process, which shape the ways in which e-governance tools are selected and can be used to support judgment.

What do policy makers do when they undertake governance? Answer – sense-making

To understand what kinds of e-governance policy makers can, do and might make use of, we need to appreciate not simply, as the typical information technology consultant or systems analyst might want, their 'business requirements'; nor indeed is it enough to secure their commitment to using new technologies, in the hope that with time, modest budgets and good will, policy makers as users will adapt new technologies to their peculiar needs, although both of these things are certainly important (McNurlin and Sprague, 1998 [1993], chs 13, 14, 17). Most important is to understand the overall nature of the task, whether or not particular policy makers can, do or would like to carry it out in particular ways, and then to understand the peculiar institutional pressures and constraints under which differently situated policy makers work. The former is the task of the present chapter, while the appreciation of institutional constraints is for Chapters 8 and 9.

There are basically three clusters of activities in which policy makers are engaged, which we may as well call gathering and organising intelligence, coming to judgment, and exercising leadership. All three of these clusters of activities are, at some level, information based, and both can be and are already to some degree supported by information technologies. In this chapter, I shall argue that they are all, in the first instance, best understood as driven by sense-making (cf. March, 1994, ch. 5; 1999b [1996]).

When gathering intelligence on public concerns, politicians and paid policy officers typically want several kinds of information. The principal categories of information sought, in any given field of public policy making, are the following:

- *awareness*: what the public knows about, recognises, understands, is familiar and comfortable with;
- *preferences*: what both the wider public and key mobilised constituencies of interests want in general value terms;
- *thought style*: how different groups of people reason and think, given particular patterns of awareness and values;
- *complaints*: formal, registered expressions of discontent with something (e.g. but not necessarily, a public service within the jurisdiction of the authority);
- *specific requests, demands*: proposals for specific changes in policy or operation by an agency within the jurisdiction of the public authority; and
- *commitments*, or the inheritance of policy, organisational capacity and political priorities.
- *trends and situation*: what the main developments are in respect of what are understood or suspected to be the key dimensions of the presenting problem;
- *relationships*: information about connections between different possible causal factors and symptoms, in order to define a problem structure;

- *comparisons and benchmarks*: information about what the problem should be compared with – for example, what might be happening if no intervention were already being made, or what experience in other cities or other countries suggests is the typical condition of the problem, or information about the past that suggests that a particular point in time is an appropriate point of comparison for the present state of affairs; and
- *effects of policies*: the likely leverage of the various tools of public policy upon the main possible causal factors, and the pressures that make for the sustainability or otherwise of such policies (for classifications of tools of public policy, see Hood, 1983; Salamon with Lund, 1989; Peters and van Nispen, 1998; 6 *et al.*, 2002; Salamon, 2002).

But it is important immediately to correct a mistaken impression that might be taken from this list. Gathering intelligence does indeed consist in part in the formal, organised, conscious activity of searching for information about the dimensions of policy problems. More typically, however, senior policy makers – ministers, chief officers, permanent secretaries – have little time for such formal, structured research work, and will tend to delegate that to others. Activities that they do themselves include, most importantly, that of *noticing* things that might be trends, clues or opportunities, making very rapid inferences from very small samples, and on that basis, trying to shift the attention of their colleagues either to or away from what they think they have detected. This, for example, is Mayor O'Malley's strategy for his interventions in the CitiStat meetings. Even when senior policy makers do engage in formal listening, seeing, searching, they are not usually trying to do so in the structured way that social scientists might do so in the course of formal empirical research. Rather, they are working in a way more akin to that of the investigative journalist or the detective. The aim of their 'noticing work' is to identify clues that will enable them to tell stories about the way things are moving.

Organising this experience is difficult, and it is not done by producing highly structured data sets, but by processes of interpretation. Those processes are, despite appearances and even aspirations, generally collective processes. Many of the conversations, meetings, formal debates and public speeches that politicians engage in and make, however casual they might seem, are often, in their most important role, try-outs and experiments in interpretation to see how far colleagues, organisations such as businesses, movements, individual citizens recognise them and respond to them in ways that meet the policy maker's objective. As these reactions are elicited, these interpretations are recursively adapted and reframed. The series of meetings conducted between the Department of Energy and other departments during 'battles of the models' provide good examples of this: in each, policy makers try out successive interpretations to see how far coalitions can be built around them, and then refine them. However, for reasons that we shall explore in Chapter 8, there are often good reasons why politicians in particular (unlike salaried analysts) do not want to create collective resources out of what they produce collectively.

The face-to-face element of this work of noticing and trading provisional interpretations is unlikely to disappear, or be wholly replaced by online work, although of course online work is becoming much more important. Quite simply, the politician, as policy maker often needs the rapid 'bandwidth' of being able to notice not only facial expressions that could be captured in videoconferencing, but the whole atmosphere, the body language, the reactions of third and fourth parties who are physically present. It is true that even very narrow bandwidth text-based communications such as e-mail systems in fact support a surprisingly great amount of tacit and implicit 'additional' information over and above what might be palpable to a naïve reader of the text (Lee, 1994; Romm, 1999). But this is possible, typically only because the communication system is embedded in the wider organisational life of relationships between the message writers, many of which will have to be face-to-face if the e-mail etiquette is to sustain a rich, tacit or implicit burden of information (Romm, 1999).

Coming to judgment is a cluster of activities that will be discussed in more detail in the next section of this chapter. However, this certainly encompasses making decisions, although, as I shall argue in the next section, in fact decision making is a derived output of coming to judgment, not necessarily the main point. A central debate about e-governance, and which will be addressed in this book, is around the question, how far can politicians and senior decision makers in government – implicitly, by contrast with board members, chief executives and senior managers in the business world – be expected to make use of, or even rely on, formal decision support systems in policy making?

Exercising leadership is also in part an activity of story telling (Denhardt, 1993; Gardner, 1997). Mayor O'Malley's tales are essentially simple ones of villainy and heroism in carrying out programmes. By contrast, the stories produced by leaders involved in the development of energy policies are more complex, including narratives of subtle compromise and noble patience over years. Leadership comprises at the very least offering interpretations of problems and issues, showing certain kinds of emotion – certainly concern, sympathy and care, but also emotions such as anger, indignation, fear, resentment, courage, and so on (Frost and Robinson, 1999) – while engaging in advocacy, persuading, shouldering responsibility for difficulties. Leadership is in part the attempt to secure some degree of institutionalisation and fixity for frames and interpretations of voters' experiences and concerns, as against those offered by rivals. The challenge for leaders is always to be seen to be at once sufficiently a taker of information and commitments from those whom they appeal to secure in order to be to some degree accountable to them, whilst at the same time being sufficiently a maker of new information that potential supporters have reason to believe that the policy maker can achieve some institutional momentum. There is no general answer to the question, where the balance lies: that must depend on the particular political situation, the policy problems in question, the relative power of policy makers individually and collectively vis-à-vis other information shapers such as the mass media, business, and movements ranging from long-term religious organisations to short run vehicles for particular redress (Bass, 1990).

Leadership work has long been done through information and communications technologies, and in most developed societies, political leadership has made its Faustian pact with analogue television and radio broadcasting. Skills have been finely honed for these media, and much of the anxiety felt by policy makers about the new digital media can be explained by their uncertainty about how their skills will have to change in order to tell stories and associate themselves with those stories in these media. Those scenarios that predict a future of fragmentation of mass audiences and 'narrowcasting' over many, many digital television channels, and those which predict that the predominant form of network access will be small screen media such as those of the current generation of mobile phones, are of particular concern. Of the two scenarios, if there is no other on offer, most politicians would prefer the former, but the choice is Hobson's.

What is really central to all these three clusters of activities is *sense-making*, or constructing accounts of situations and events with which to interpret actions, intentions and events, giving purposes, identifying phenomena that may be anomalous within those frameworks and so perhaps present challenges. Sense-making is an ongoing process of unending revision, undertaken subject to the constraints of prevailing institutions which press certain interpretations upon sense-makers and discourage others (Douglas and Isherwood, 1979; Louis, 1980; Douglas,

1992b; cf. March, 1994, ch. 5, 1999a and Weick, 1995, 2001). The principal outputs of what policy makers do when they gather intelligence, come to judgment and exercise leadership are interpretations, understandings, ways of framing experience. Sense-making is the building of narrative (Stone, 2002 [1988]; Fischer, 2003, ch. 8) by which experience becomes *legible*, in which the relationships between events become intelligible as causes and effects in ways that tell a story about the scope for efficacy in action or intervention, and in which the trajectory of events is given an evaluation as benign, malign, dangerous, fragile, hopeful or otherwise, for the sense-maker and their community or organisation or peers. Moreover, sense-making is guided by *accountabilities*, both those which are objective and those which are subjectively recognised (Douglas, 1980). In making sense, we describe how we want those who have shaped experience to be held to account, how we think they are in fact held to account, and how we relate to the institutions that hold to account both ourselves and those others who have shaped our experience.

A leading commentator on sense-making (Weick, 1995, ch. 2) characterises it in the following ways. It is, he suggests, in the first place *retrospective*, fitting events into a narrative sweep about where we have come from and where we are going, and most importantly, in a narrative about identity or who we are, where 'we' can mean – in the present context – the citizenry, the particular section appealed to, the political organisation, government or some other salient force. In O'Malley's Baltimore, for example, the retrospective analysis of performance is a key element of political judgment, because determining individual responsibility has been found there to provide the kind of stimulus to future action that the Mayor wants. Secondly, sense-making is *projective*. Making sense involves the framing of expectations, of possible futures, dreams and nightmares, fears and hopes. Next, it both creates and embodies – 'enacts' in both the legislative and theatrical senses of that word – the framing of experience it offers. Producing a policy makes a real change in the world: even what is wrongly dismissed as 'rhetoric' is never 'empty'. This may take the form of analytical projections and scenarios of the kinds used in the Department of Energy, or a much less formal vision.

Sense-making is a *collective* process: no single policy maker can make sense of anything. Even individualists such as O'Malley work against the foil of the sense-making of his agency managers. Sense-making is also continuous, although, as we shall see in Chapter 8, policy makers situated differently toward the sense-making process may experience the flow of time in sense-making very differently, and this will have big consequences for the kinds of e-governance tools they can most effectively use. It is based on the noticing of cues and clues and its first key function is to move and fix attention between all the competing cues and clues available.

Finally, the criteria for good sense-making is that it should seem plausible to those for whom it is made, given their particular situation, rather than that it should necessarily be strictly and rigorously accurate. Bad policy can be founded on the most rigorous policy science, and good policy can be founded on hunch, gut instinct and the astute interpretation – even misinterpretation, over-reading and mistaking – of quite casual clues.

If this is accepted, then the kinds of e-governance tools of which policy makers stand in greatest need are those which support the elements of sense-making (Weick and Meader, 1993). Unfortunately, this is not the end of the matter, for, as we shall see, when we appreciate the institutional constraints upon differently situated policy makers, it becomes clear that we cannot rush from the importance of sense-making directly to selecting kit, and many hopeful reformers in recent decades have failed by attempting just this. However, because of its peculiar importance in politics and public policy making, it is necessary first to explore what is required in political and policy judgment.

Political judgment

Chapter 1 suggested that the key activity of policy making is that of political judgment. In the previous two sections of this chapter, it has become clear that this does not mean that this skill consists in a facility simply to make good decisions, according to some criterion or other. I have argued that a key weakness of some of the most common ways of thinking about what policy makers actually do, is that they lead one to imagine that their basic activity is the taking of decisions. For example, most conventional studies of judgment focus on the immediate moment of decision and consider it as an optimisation problem, and then examine deviations from optimal rationality in the assessment of probabilities or the selection and representation of options (e.g. Kahneman *et al.*, 1982; Dowie and Elstein, 1988; Kahneman and Tversky, 2000). This approach is flawed in several ways. First, it is highly normative, starting from where people 'ought' to be, from a single intellectual standpoint, and then maps deviations. Second, the approach is interested mainly in average or typical 'deviations': few attempts are made in that tradition to develop typologies of syndromes of rationality. Third, such models ignore the central importance of the *context* – specifically the context of the social and political organisation of the decision makers – which explains why some options and some probabilities are represented in the way that they are. Fourth, these models assume that good decision making is focused exclusively on the technical aspects of the problem. A genuinely political approach to understanding policy making must appreciate that tackling a policy problem in the manner that it has to be represented to the voting public may be but one element in the considerations that must weigh with policy makers. Fifth, these models miss what is crucial to *political* judgment – namely, its character as something negotiated between differently situated actors, argued, rhetorically mobilised and grounded (for discussion of this feature in the context of e-governance, see Lenk *et al.*, 2002).

Rather, I have argued, decision making flows out of the larger and very differently demanding activity of making sense of events, processes, situations, problems, and the shared life of the communities that elect or sustain them, in political and policy terms. The point is not that making policy decision is somehow not very important, or that e-governance technologies that support decision making are somehow irrelevant. Rather, as we have seen, the point is that e-governance technologies that support sense-making are of very great importance, and the use of decision support systems is most sensibly embedded in the larger context of support for sense-making by policy makers (Weick and Meader, 1993).

However, such systems must to some minimum degree rely upon policy makers' possession of, and to a further degree, help them to sustain, cultivate and support certain skills of sense-making and, flowing from this, decision making. Indeed, a central criterion by which policy makers should, it follows, appraise e-governance systems is the extent to which they can make those systems serve this end.

Judgment, understood in a richer and fuller sense than that in which it is nowadays commonly used, is the central faculty in sense-making as a whole (cf. Denhardt, 1993). Typically, the word 'judgment' is used narrowly in many texts on risk perception and decision making to mean the capabilities of calculative reasoning upon available evidence in order to come to a decision between already defined alternative options: for example, Kahneman *et al.*, 1982; Hastie and Dawes, 2001; or at most a limited set of cognitive strategies for competent decision making, for example, Hammond (1996). Studies on political judgment mostly focus on astuteness in appreciation (Renshon and Larson, 2002) or else on the question of the relative weight of intuitive and analytical strategies

(Beiner, 1983; Steinberger, 1993). There is also a large literature in political psychology under the rubric of 'political judgment' which is not so much concerned with decision making as with popular political attitudes (e.g., Elkin and Soltan, 1999; Marcus *et al.*, 2000).

These studies are, in the present usage, concerned with only some of the elements, and not necessarily the most fundamental element in policy and political judgment as it is understood here. This book is concerned with processes of coming to policy and political judgment about courses of action and about institutions in organisations with public authority accountable to senior elected officials and through them and through the media to the voting and taxpaying public. It will be argued that the literature building on the work of Hannah Arendt (1954a,b, 1992a,b) in political philosophy is more helpful (Beiner, 1983, 1992; Yar, 2000; Beiner and Nedelsky, 2001).

Arendt stressed that political judgment is a collective activity conducted in deliberation, as part of a community of judgment, not an individual capability of leaders or would-be leaders or individual citizens making judgments like armchair generals: even solitary political reflection requires, she argued (1954b), the representation of the viewpoints of others, and the particular others to whom we are bound and before whom we judge and define the nature of judgment, whether it is moral or political (Beiner, 1992, 113). Moreover, Arendt's view of technocratic and bureaucratic forms of decision making should show the critical reflection that is of the essence of judgment (1992a [1958]). It is because she defined political judgment as critical reflection and deliberation either with others or at least by people who will be accountable to others with whom they live under at least some common institutions, that Arendt saw political judgment as an activity which is good in itself, irrespective of the merits of particular reasoning or decisions that flow from it (Beiner, 1983, 1992).

A second source of richer understanding can be found in the analysis of policy judgment by Sir Geoffrey Vickers (1995 [1965]). Vickers, a career policy maker tuned social scientist in retirement, developed a sophisticated account of policy judgment as grounded in what he called 'appreciation' or sense-making about the context of decision, grounded in shared practice, which combines recognition of both facts and values. For both Arendt and Vickers, orientation to different points in time is crucial to their anatomy of the distinct components of judgment, and this will be central to the account that will be offered here.

In this section, I offer a characterisation – not, strictly, a definition – of political judgment, in order to show what kinds of capabilities need to be both brought to the use of e-governance tools and cultivated in their use, and at the same time, the focus of design for those developing systems for policy makers to use. In Chapters 8 and 9, this account will be enriched. However this characterisation of political judgment provides the foundation for the principal argument of the book. For, as noted above, a central finding to emerge from the review in Chapters 2 and 3 of the tools of e-governance is that what makes for effective e-governance is not so much the technical quality of the software design, as the situating of the use of that software in the mesh of social relationships between policy makers. I shall show that political judgment is not principally an individual faculty at all, but an achievement of the social and political structure of relationships between policy makers. In Chapters 8 and 9, I shall present a way of classifying and understanding that social basis of the effective use of information technologies in governance, which will then enable the development, also in Chapters 8 and 9, of a still richer conception of political judgment and the roles that such technologies can be made to play.

Drawing on accounts from a range of disciplines, I will distinguish five crucial elements or moments in political judgment. These are not 'stages', in the sense that they are typically or ought ideally to be gone through in sequence, or that none can be missed out. Rather, they are moments or elements of judgment in an intelligent performance by all the groups of policy makers who have some bearing or influence upon a field of policy, a problem, a crisis. I distinguish them by their relationship to time. For when policy makers make sense of the conditions and problems they face, political time and policy time are crucial factors. The time horizons over which the problems change – environmental shifts, the business cycle, the electoral cycle, the educational career of children, the period before measures can be expected to produce measurable results or whatever – are basic structuring features of the sense that policy makers can make of the relationship between their levers of power and their problems and limitations. So too is the time available to them for making sense and for acting or enacting. Making sense of events and contexts in ways that support decision making necessarily involves the development of narratives about past time. As we shall see in later chapters, the basic institutional settings in which judgment is exercised differ crucially in their support for sense-making over long and short collective memories of the past and collective future planning horizons (see also 6, forthcoming).

The first element is essentially oriented to the *immediate*. This is concerned with managing the problem of what to pay attention to. For the scarcest resource in any organisation, and this is particularly sharply true in the context of public governance, is always the attention of policy makers. There are so many competing demands for attention, and so many rival criteria of what counts as important, than the preliminary question of what matters are worthy of attention is the most fundamental (March and Olsen, 1976; March and Simon, 1993 [1958]; Weick, 1995, 2001; Simon, 1997 [1947]). The first element of judgment, then, calls for astuteness in very rapidly balancing the competing demands and rival criteria. A finance minister, for example, must allocate her attention by taking some account of what powerful lobbies are demanding, what the existing overall strategy and commitments are, what their long-term goals (if any) are, the shocks and events of the moment and her informed view of their significance, the pressures from spending ministers, the likely consequences of her changing priorities for the bonds, gilts and currency markets and so on. By astuteness here, I mean a compression of all the other elements to be described in a moment. However, this astuteness is essentially oriented to a rapid purview of the immediate pressures, which enables the other capability of sense-making – namely, the exercise of imagination to identify opportunities, risks, strategies or at least tactics. Yet astute choice is constrained by criteria that are as disciplined by practicality, by leaving open as many options for the future as one can while making the necessary commitments, as they are by moral principle. Astuteness is not moral heroism in politics, but neither is it not mere adroit manoeuvre or 'politicking', for it is concerned with the balancing of multiple criteria for appraisal, and it must fore-acknowledge public responsibility in ways that the merely adroit intriguer does not.[6]

Moreover, astuteness is not an individual characteristic, but a collective achievement (contrary to writers such as Fleischacker, 1999). Politicians and senior officers learn from each other what to look for, what to notice, what counts as intelligent balancing of criteria, and they do so iteratively over a series of overlapping constituencies in meetings, conversations, joint projects and so on. Much of that learning is tacit, highly specific to the particular group and dependent on the particular context. A group of decision makers who have worked together over any length of time exhibit informal norms and institutions around sequences of speech, shared vocabulary, division of intellectual labour, that encode their collective astuteness, in which each person's contribution is dependent for its continued functioning on the presence of the others. It is this dependence of the collective astuteness on the present continuation of the group that enables such astuteness to be attuned to judgment of the immediate.

Second, there is a vital element of judgment which is *backward-looking*. A first element of this is sometimes called 'appreciation' (Vickers, 1995 [1965]). The first part of this – the 'reality judgment' (Vickers, 1995 [1965]) is that aspect of sense-making which requires the gathering of information and intelligence about the causal forces that have shaped the crisis, the problem, the field, the opportunity. How far back the exercise of backward-looking judgment can or should go will vary with the extent and the style of organisational memory (6, forthcoming). For example, Mayor O'Malley was largely unconcerned that the CitiStat system enabled his analysts to look no further back than the year he came into office. By contrast, the Department of Energy uses data going back many years for its modelling and analysis work: determining the date for retrodiction is a crucial political weapon in the modeller's armamentarium.

In backward-looking judgment, there is a moment of imagination, in which the policy maker – or rather, the policy making groups and communities – must compare the many ambiguous or equivocal ways in which the issue can be framed (Weick, 1995, 2001), and also think counterfactually, about what might the appropriate benchmark for comparison with the actual pattern of processes and events be.

There is an unavoidable normative element in the backward-looking judgment – namely the allocation of responsibility not only for the creation of a problem or condition, but also for any actions in response to it. An individualistic variant of this moment of judgment marks out Baltimore City's CitiStat programme: the retrospective assessment of organisational performance is expressed through the quasi-judicial appraisal of the performance of the individual manager. Arendt stressed the importance of judgment of historical context and of the responsibility of individuals and groups, not only in the judicial context but in wider political debate, as vital to the cultivation of political judgment generally. It is this that is the real subject matter of *Eichmann in Jerusalem* (Arendt, 1963): she saw the international debate about the extent and nature of Eichmann's real responsibility as an example of a process that is vital to the functioning of a decent polity able to balance the *vita activa* and the *vita contemplativa*. By contrast, she saw in Eichmann's own inability to reflect upon his own life and work with backward-looking judgment as his deepest and most significant moral failing, and one that had been in some part institutionally produced. Backward-looking judgment comprises three basic activities. Normative retrospective judgment typically takes the form of the allocation of blame or praise, despite the common protestation in many administrative contexts that blame is not at issue or not appropriate: Eichmann represented the terrifying logical extreme case of that syndrome. In fact, as Arendt rightly argued, blame and praise are essential to appreciation of the moral texture of any policy issue: although (as we shall see in Chapters 8 and 9) some institutional styles are more concerned with confrontative, punitive and unforgiving styles of blame than others, the appreciation of responsibility is always central (Douglas, 1992a). One cannot be said to have developed an appreciation, of a kind that would be adequate for political judgment, of crime, global climate change, declining competitiveness or a collapsing exchange rate, until the move has been made from identifying causes to appreciating the responsibility of different agents for those causes.

Third is a moment of *process-oriented* judgment. That is, this element is oriented to the ongoing process, not just the immediate moment. One part of this is sometimes called the 'instrumental judgment' (Vickers, 1995 [1965]), for it is concerned with the imagining of options, the lateral thought that brings together general ideas about ways forward, strategies, coping mechanisms or solutions are designed for policy issues with the particular characteristics of the present problem or condition. The priority given by Fairfax County to the use of the Group Decision Support Centre shows how important this element of judgment is in that authority. It involves both the consideration of how such actions have been used in the past and may currently be in place, as well as projecting them forward together with projections forward of the problem itself (which are taken from the appreciation). But this is not all. For judgment is not a technical procedure for taking decisions about defined problems. The options identified must be set against the larger process of making sense of experience that policy makers are engaged in: for example, redistricting in Fairfax was not an optimisation problem, but a highly political one for which the GDSC and GIS tools provided the arena only. Options generated must be situated in – or found to be incapable of being situated in – the larger narrative about the style of politics, policy and governance exercised, the policy makers' conception of the world and the nature of their problems, and so on. In short, the imagination of candidate choices must be situated in the overall narrative to which policy makers are working.

Fourth, there is an element of *forward-looking* judgment. The distinction between backward and forward looking moments in judgment corresponds to the two conceptions of judgment in the thought of Hannah Arendt. The first conception is focused on forward looking judgment for action (Arendt, 1954a,b), or what she called the *vita activa* (Arendt, 1992a [1958]). The spectator-oriented conception is focused more on backward-looking judgment of responsibility (Arendt, 1992b: see D'Entrèves, 1994; Yar, 2000). The two conceptions are brought together in Arendt's (1963) central claim that without the capacity to institutionalise in a polity a process of critical judgment of the past, there is little hope of cultivating capacities for decent and intelligent forward-looking judgment.

Under the rubric of forward-looking judgment, there has to be undertaken a process of deliberation, in which policy makers attempt to give themselves some provisional distance from the options, in order to see just what commitments, and which groups in the society, economy and polity might benefit from them, and then try to work out how they might balance all the considerations for and against options and combinations of options, to reduce the numbers of options and to identify the trade-offs between risks, costs and benefits. It is this which is to some degree addressed by conventional decision theory. However, forward-looking judgment requires a social context and process about which conventional decision theory has nothing much to say. Distancing is not simply a matter of trying to detach oneself from special interests, but also a process of trying to stand back from the particularities of the situation, and to see it as part of the larger activity of making sense of political experience – what politicians love to call 'the big picture'. This is not achieved by way of individual cognitive bootstrapping but only in the course of certain kinds of meetings, conversations and collective uses of intellectual disciplines in which people hold one another to account for the quality of their distancing. This necessarily involves certain kinds of ritual activity in those meetings. The ritual order used, for example, in Fairfax county's group decision support centre is one in which the physical space of the room, the allocation of roles between participants and facilitators, and the manner in which the technology is designed to limit certain kinds of expression and encourage others, work together to produce a ritual style of deliberation which may not achieve the required distancing in every individual, but which is hoped and believed to increase the probabilities of achieving adequate – but of course never complete – distancing among the group of decision makers required for the purpose in hand, given their background of appropriate prior commitments, roles, interest and institutions which are not to be the subject of distancing.

Next, there is a moment of coming to closure, exercising choice, and making a commitment. A choice made requires that first the option or options committed to are then crafted, refined, or tailored: that is, judgment calls for a process of design to develop and work up a preferred option. In particular, this crafting is done best when strategies are crafted to provide future opportunities for judgment, by ensuring that options chosen are ones that yield information for backward-looking judgment or appreciation in future, and – most importantly – are the ones that will alert and focus policy makers' attention when and only when appropriate. This too is a collective process of refinement in the course of deliberation, through a series of try-outs.

Fifth is a complex moment of combined forward- *and* backward-looking judgment. This is an essentially normative moment, in which policy makers specifically take *responsibility* for the sense-making they engage in, by taking ownership of the foreseen but unwelcome consequences of their commitments and their narratives, and accept the risks and costs implied in the trade-offs they make in settling upon a course of action. Yankelovich (1991) emphasises that in public opinion research, an attitude should not be dignified with the title of a judgment unless those who profess to hold it show in response to further questions that they do understand and accept responsibility for the costs and risks of their preference. Here, the appropriate tense for judgment is the subjunctive pluperfect: the policy maker considers what she *will have been* held responsible for. Rarely is this the product of solitary reflection, or even if it is, then that reflection will be conducted against the background of prior and subsequent accountabilities to fellow decision makers, particular interests, perhaps intellectual disciplines of analysis and decision making and to wider publics which will instil that recognition of responsibility.

In Baltimore, for example, Mayor O'Malley believes that he will and should be judged in the relatively near term on the impact that CitiStat has upon the quality, responsiveness, efficiency and timeliness of the city government's services, and he uses his role in CitiStat in ways that demonstrate his sense of ownership for the means he has chosen in the light of those ends. By contrast, the Department of Energy staff who have crafted the concept of the 'responsible chart' are recognising that they will be judged over the long term by citizens, their own profession and their successors on their responsible custodianship of the professional ethics of the analyst as independent public servant trying, always in highly straitened political circumstances, to balance serving the elected politicians and speaking truth to power. Both are thinking in the subjunctive pluperfect, but in the time horizons of their institutional settings.

Each of these five dimensions of political judgment acts upon the others, to produce a process of making both political and policy sense of a situation, a problem, a crisis. Taking responsibility, for example, is a way of allocating attention to particular risks, costs, opportunities and benefits, and directing the political organisation to gather certain kinds of information and at least provisionally frame problems in particular ways, which will shape future understandings and particular judgments. Narration too is a way of situating both attention and of identifying what is available to be taken responsibility for and what can feasibly and acceptably be sloughed off to others.

Box 7-a summarises these five elements of political judgment.

Box 7-a   Political judgment

1. Immediately-oriented judgment.
Priority attention: allocating attention intelligently.

2. Backward-oriented judgment or appreciation.
Reality judgment: gathering information and intelligence especially on causation.
Imagination: structuring problems, imagining rival framings, appreciation of variety and variation.
Allocating responsibility: assigning blame and praise for causation and responsibility for action.

3. Process-oriented judgment.
Instrumental judgment: imagining options, projecting, appraising feasibility.
Narration: fitting choices to larger sense-making, narration, commitments.

4. Forward-looking judgment.
Deliberation: distancing from options and advocates, balancing considerations.
Valuation: coming to closure, choosing, making commitment.
Design: crafting particular strategies, crafting future judgment capabilities into them.

5. Forward-looking and backward-oriented judgment.
Normative: taking responsibility for commitment, for less welcome consequences, for trade-offs.

The five dimensions distinguished here are described in cognitive terms, which is appropriate in describing what is in effect the dependent variable in the argument. Yet political judgment, as it is understood in this book, is not a practice, a skill or a virtue than an individual politician, advisor, policy analyst or chief officer can alone, cultivate or exercise, still less 'possess' in the way that someone can be said to possess a discrete piece of their own knowledge or a belief. It will be argued that it is rather something essentially conducted and achieved, to the degree that it is achieved at all, by the whole political organisation – the Cabinet, the senior management team, their advisors and staff, the governing party, under the scrutiny of the legislature, media, interest groups and many wider publics. It is, almost by definition, a collective enterprise, although not necessarily or typically one in which all the contributors are equal. Political judgment as sense-making is the political and the policy intelligence of the organisation committed to governance. But that intelligence is, as we shall see in Chapters 8 and 9, a process shaped by the encounter between different and potentially conflicting institutional situations. The argument of the next few chapters will be that the relative weight of each of the five moments or dimensions in political judgment is the product of these collective and institutional processes, and that the case studies presented in Chapter 6 illustrate this well. In particular, Chapters 8 and 9 will argue that the distinctive styles of experience of time by people differently institutionally situated in the policy process will bias them toward those e-governance tools that they can best use, and which will thereby give different emphasis to each of these five moments of political judgment.

8
The Organisation of Policy Makers and the Use of E-governance Tools

This chapter offers an account of the settings in which different kinds of policy makers find or put themselves, which, I shall argue, is the fundamental shaper of their willingness to use different kinds of e-governance tools, and of their ability to do so successfully to sustain political judgment. The chapter begins by distinguishing four typical institutional settings at the level of ideal types but in fairly empirical terms, before going on to show on theoretical grounds that the basic characteristics of the four ideal typical cases are indeed powerfully illustrative of the range of situations in which policy makers work, quite generally.[7] At least three of these four can be readily discerned in the styles of political judgment sustained with the use of technology in the three case study authorities discussed in Chapter 6. The argument then moves to exploring the kind of institutional styles of or biases in exercising demand for information, use available information, cultivating and exercising judgment, learning and decision making that can emerge in each of these four basic settings. This analysis makes it possible, in the following Chapter 9 to look at the kinds of e-governance system that will most readily appeal, and that will most readily be rejected, in each of these four basic situations.

The normative implication of argument of both this chapter and Chapter 9 is that we should give due recognition to and place for each of the four settings. None can be eliminated from policy making, nor should e-governance tools be designed or advocated on the basis that only one of these settings is desirable, still less in the hope that using technology to support judgment can promote one setting and its bias at the expense of all the others. In Chapter 9, this approach will provide the basis for developing ways forward.

It is critical to understand what politicians and senior policy staff want from e-governance technologies, what they can make use of, and how they use such systems. Most of the mistakes made in the history of computer support for policy work have arisen from failure to do this. In some crucial respects, politicians and senior public servants are no different from other organisation leaders. It has long been recognised that the relationships between decision makers, the way they are organised in relation to one another, the social context in which they deal with one another and with other organisations and their wider publics, are at least as important in shaping what they will believe and choose, as are any characteristics of the problem they are addressing (see e.g. such classics as Gulick and Urwick, 1937; Barnard, 1938a; Simon, 1997 [1945]). Moreover, they are much more important in shaping actual decisions made, a now vast body of research suggests, than is the technology used to support those decisions (Shapiro *et al.*, 1991; Jones, 1994; Bijker and Law, 1994; Weick, 1995, 2001; Bannon, 1997; Kling, 1997; Hayes and Walsham, 2000), or the actual information policy makers possess about problems and solutions (e.g. Weiss, 1979, Feldman, 1989; Scott, 1992; Overman and Loraine, 1994; Heinrich 1999). As many decades of work in the sociology of knowledge has shown (Berger and Luckman, 1966, Schutz, 1967, Douglas, 1972, 1982, 1986; Mannheim, 1985 [1936], Durkheim, 1995 [1912a], Zerubavel, 1997; 6, 2003), how people are situated in relation to each other shapes what they can think, know and decide: these social forces have a constraining power that are properly called 'institutional' because they are more or less entrenched and rule-like social constraints defining structures of accountability (Durkheim, 1982 [1895a], Douglas, 1986). I shall deploy a particular application of this general argument to show how these forces will shape the technologies of e-governance that policy makers are likely to find most congenial, the ways in which they are most likely to make use of those systems, the styles of decision procedure they are most likely to find appropriate and the aspirations and anxieties they will most likely bring to bear upon considering the prospect of greater investment in and use of e-governance tools.

Four institutional settings in policy making

The heart of the argument here is that we can classify the basic situations in which policy makers work and that shape their interest and capability for using the tools of e-governance in a rather parsimonious way, which will help us to understand their potential and also their limitations.

The snap! players: backbenchers and stranded policy staff

Consider first the situation of many backbench Members of Parliament in a Westminster-style system or of many backbench local councillors, in large party groups where there is strong party discipline exercised by whips and where the party has strong leaders. Indeed, in stable two party systems under electoral systems (such as 'first past the post' still used in Britain for general and many local elections, although not for the Scottish, Welsh or London assemblies) that militate against fragmentation and reinforce majorities and party discipline, these are probably the most frequently observed conditions of party organisation. The whips hold backbenchers to account, and only loosely and in the long run do backbenchers, when they organise rebelliously and differently from their usual situation, hold whips to account. For backbenchers are often regarded as, and behave as, little more than vote fodder by the party leadership. Promotion opportunities are few, and seem to be dependent upon the caprice and favouritism of a leadership that can seem to know little about the real work and capabilities of backbenchers. One route to influence is through faction and cabal, but where discipline is strong, this can actually seem counter-productive. Most backbenchers, however assiduous their constituency work, have limited ties to their constituency, and face many incentives to represent their party to their constituents more often than to advocate constituency interests within the party. Their links to one another are weak: there are few powerful reasons and few organisational structures through which to collaborate with one's peers. The backbencher is not without influence, but opportunities for influence – though real – seem to come rather randomly, when chinks open up in the party leadership, when formal consultations are suddenly announced, in free votes, when the chance of rebellion occurs without completely wrecking all hopes of promotion, as may occur when the governing party in a parliamentary system has a wafer-thin majority. The situation of the backbencher is at once highly constrained by the authority of the whips, and on the other hand, constrained more subtly by the weak bonds she has to other policy makers and even fellow backbenchers. The resulting randomness of her political experience is characteristic of such situations, at once regulated and isolated. (By using the term 'isolate' to describe the situation, I do not mean that such people *know* very few people as, say, acquaintances: on the contrary, they may have a very larger acquaintance, and many backbenchers do: rather, the comparative isolation is a term capturing the condition of their *accountabilities*, which are weakly connected with other in their peer group but strongly regulated by powerful institutions.)

The backbencher in this situation possesses very little valuable policy relevant information that is not in any case in the public domain. While there is no particular reason not to divulge the information they possess to other backbenchers, there are few regular or structured occasions to do so, and few great benefits from doing so. The incentives for specialisation, or the development of bodies of information and expertise in particular policy fields, are not very strong for backbenchers: some will find reasons to do so, but others will not. In general, a more eclectic style of information and intelligence gathering is more likely to serve their situation in order to make the best use of the opportunities available to them. Information and knowledge are, for the backbencher in this situation, casually available resources.

In this critical respect, the situation of the backbencher in Westminster-type parliament is not unlike that of many others with some stake in the policy process. There are some analogies within the civil service. Heads of small agencies, disregarded by the permanent secretaries of the departments to which they report, with limited budgets, few connections with other agencies, save when called in like a plumber to fix a particular problem, face a situation not dissimilar in its highly constrained, weakly bonded structure.

In presidential and congressional systems such as in the United States, where the right of legislative initiative is not monopolised by the executive and where party whipping is a much less powerful institution and where committees represent major opportunities for securing and exerting influence, the position of the individual members of the House of Senate is, of course, very different from that of the backbencher in a Westminster-style parliament. That is not to say that highly regulated and rather isolated individuals are not be found in the policy process in United States or in similar states. More often, however, they are to be found within the career executive in policy roles within departments, even at quite senior levels. In the Department of Energy, for example, we saw in Chapter 6 that in the final year or two of an administration, the position of the individual senior policy staffer can become very precarious, if others might consider that she or he has been too closely involved with the incumbent administration and if a change of party control is expected for the forthcoming presidential elections. In this situation, trust between colleagues is eroded, and the staffer faces the possibility of being moved sideways, shunted into much less influential roles, even pushed into early retirement. Their options are few. They can gradually disengage from their policy commitments, engage in more defensive policy advice and development, and draw less attention to themselves in the hope of surviving. Alternatively, they can accept the likely fate of being sidelined. Opportunities for collective action decline for a staffer in this position.

Like the backbencher in a Westminster-style parliament, though, the isolated vulnerable policy staffer in a senior staff role at the end of an administration for which they have worked assiduously is by no means unimportant in the policy process. People in these positions make important contributions, at least in the aggregate, and indeed their defensive, coping strategies do much to define the terrain upon which others in the policy process must work.

The game that policy makers in this situation must play with information is much one the card game, snap! The cards are dealt by others, opportunities seem to come at random, and the game has little structure.

The poker players: the cadre of leadership candidates and the administrative barons

Compare this with the situation within the leadership cadre of a political party, and especially of one in office, either nationally or in a major local authority. A national party leader or the leader of a party group in a city or county council is surrounded by people whose loyalty cannot be indefinitely relied upon. The more senior are her lieutenants, the more ambitious they are, and the more they want her job, and the more capable they are of overthrowing her, should she fail them electorally, make mistakes that cost them public opinion poll ratings, or should she fail to consult them adequately. The cadre comprises people who are rivals, and exhibit varying degrees of short- and long-term ruthlessness in pursuit of their ambition. For without that driving ambition, they would not have survived long enough to make their way from the back-benches to become a leader. Indeed, if they were not initially ambitious for the highest office when they began, the taste for access to real influence, and the fear of loss of role and face make ambition the only way to survive, having once entered the charmed circle of leadership. Competition must be waged through the policy process, and what may often be most important about their input into policy making is that their personal signature is publicly associated with a policy that is seen to be popular, or at least effective even if unloved. This individual brand-ing will typically matter much more than the technical soundness of the policy design.

Executive politicians in this situation cannot afford to trust each other very greatly. They will know that their peers, however collegiate their blandishments, will always be rivals for budgetary resources, for influ-ence over policy, for the favour of the current leader, for access to the media. In this situation, calculation and strategy is all that matters. Information is valued as long as one's own control over it can be secured, and as long as the fact of one's own possession of it is not generally known.

Cadre politicians and senior career public managers both have strong incentives to specialise (although they may need to change between spe-cialisms rather quickly), because in this way they can consolidate their authority over their own domain – principally, a department – and they can control information flows to and from their peers within the cadre. Their position gives them access to plenty of policy relevant informa-tion that is proprietary to their departments, or even to their own offices. All the incentives they face, however, are to keep this private, and not to trust other members of the leadership cadre with more of this information than they need to, or than is useful in securing temporary alliances around the cabinet or the executive group table. In particular, the information must be kept private on the basis of which others could reconstruct one's fall back position in budget negotiations, in negotia-tions on joint policies and operations, in relation to support in the wider party, and so on. Information and knowledge are, for the leadership contenders, essentially ammunition for advocacy and defence.

If the game that backbenchers play with information seems like a kind of snap!, governed by the random shuffling of cards by others who act as the dealers, the game played by the senior executive politicians with their information is hard-faced poker for high stakes. No one shows their cards to anyone else; bluff and counter bluff are the essence of negotiating strategy; vengeance is swift and effective, and there is no mercy for the fallen.

By contrast with the backbencher's situation, the situation of the member of the leadership cadre is relatively unconstrained by the authority of the party whips or even the leader, for she can manoeuvre within the leadership group to influence the exercise of discipline over the backbenchers and the ordinary members, and, if things go badly wrong, she can quietly allow selective 'leaking' of information against others in the leadership cadre, should it benefit her to do so. She has much more access to the levers of influence than backbenchers, and so is less constrained by the authority of the party line, for she can not only determine what that line is, but influence – or at least re-describe – what counts as conforming to it. Yet, curiously like the backbencher, her bonds to her colleagues are – also rather weak. This is perhaps surprising, given the official importance of the cabinet's collective responsibility and the apparent importance of party unity and being seen to 'hang together or else hang separately'. Again, not wholly unlike the way in which the backbencher experiences political life and the policy process as somehow random, for the member of the leadership cadre, there is an unpredictability to the experience of power. Unlike the backbencher, the minister or local executive figure experiences power rather than access and influence, and therefore the unpredictability is something over which they have some influence as a player, and therefore, the randomness is not something 'given' as for the backbencher. Rather, it is the effect of the kind of strategic interaction between informed, Machiavellian individuals that is well modelled in game theory. In short, this is a situation of limited constraint and also of weak bonds both to peers and subordinates. This is the one that places accountability demands upon those within the cadre which are most individualised, and most likely to take the form of *ex post* accountability for performance and least likely to take the form of *ex ante* instructions on activities or the allocation of resources in any detail: in this sense, it is appropriate to speak of the accountabilities as limited or importantly 'weaker' than those faced by others in different settings.

In congressional-style systems where legislators are not subjected to powerful whipping by strong party machines and where there is strict separation between legislature and executive, at least in strong and powerful legislatures, the position of the elected senator or representative is probably more akin to that of the poker-player than the snap! player, although in systems where the executive is dominant, some kind of hybrid of these two might be expected to describe the legislator's situation.

Again, when we look across the policy process, it is not only among elected politicians that this game of poker must be played. The same characterisation can be given to the situation of permanent secretaries, of chief officers who head the major departments in a big local authority, some heads of major regulatory agencies: again, of course, there are important differences within these groups (for some evidence on differences in the positions of chief executives in local authorities in several developed countries, as measured by network position and by policy advice role, see e.g. Mouritzen and Svara, 2002), but at the ideal typical level, some important and commonly found features of their situation are well described by this institutional form. The style of negotiation between local authority chief officers heading major spending departments in the run up to the setting of the budget by the council members is no different. Information is either hoarded or given sparingly to 'colleagues' on a need-to-know basis where barter is worthwhile for both parties. Complex fallback positions are developed and only revealed when they must be used. Even the relationship between a chief officer and the elected member who chairs the executive committee that oversees her department can be Machiavellian. The chief officer possesses the information advantage over the executive committee chair, for she has the capacity to edit the material that is made available to members, and if she cannot trust her committee chair, she can control that information very tightly, even – in necessity, when the department's vital interests are at stake – quietly briefing her committee chair's rivals with chosen pieces of information.

Mayor O'Malley's Baltimore shows what can happen when a dominant cadre politician is determined to use his own position as the player holding the strongest claims to information to force everyone else to reveal theirs, and thereby weaken their competitive position, and control the extent to which there can be a competitive game of poker within the city government. The central purpose of the CitiStat programme is revealed by the language the analysts used – that the process leaves agency chiefs with 'no place to hide', and that 'a map knows no loyalty'. In essence, CitiStat's role is to reinforce the very individualism of accountability which is the hallmark of the poker game, whilst making sure, as far as possible, that there is one winner and that the other would-be players hold very few proprietary pieces of information.

The chess players: professional central policy unit staffers

Next, consider the situation of the professional policy analysis unit within a ministry, in a chief executive's office in a local authority, or working directly with a minister, in the centre of national government such as the Cabinet Office in Britain or in the Prime Minister's department in Spain.

Such work is highly structured in several ways. First, issues are only allocated for the expensive treatment of professional policy analysts when politicians or chief officers consider that the results are likely to be worth the time and effort.

Second, some of these professionals work within the intellectual disciplines of the policy sciences, and they know that the policy papers they write or the advice they give must be defensible in those terms to the community of media pundits, academics, think tank specialists, consultancy firm experts and other 'wonks' among whom the arguments they offer will be debated. For example, a justification for a policy of deregulation or reregulation of some phenomenon (price, quality, safety, etc.) in some industry may be offered – and typically they will know that they must justify the policy preferences of their minister or at least of their department – but it can only be offered in terms that fellow policy professionals will recognise. It must compare the relative scales and probabilities of both market failure and government failure, it must assess the scope of industry externalities and the risk of regulatory capture by the regulated industry; the argument must ground itself in account of the incentive structure facing producers and the degree of consumer power; it must present a credible analysis of the scale and possible management of regulatory information asymmetry and of the dilemmas of deterrence and compliance strategies in regulation; it must consider the public opinion data on the legitimacy of regulatory intervention of various kinds; and so on. Information and knowledge are, for the central professional unit, analytical tools and procedures.

Third, the work is structured by the imperative for specialisation, typically by policy field but also by method. The distinctive role of the EIA in the Department of Energy provides a good example of specialisation by method as well as by field – in that case, an agency dedicated to quantitative modelling in the energy field. Across British government, for example, there are horizontal collegial structures for economists, social researchers and statisticians, to sustain their methodological specia-lisation as well as the vertical structures to sustain specialisation by policy field both within departments and within central policy units. Specialisation, in this setting, however, is the vehicle for the sharing of information within the team, as well as also serving the secondary function of securing the hoarding of information and keeping it from other departments with conflicting or competing interests.

These three structuring forces give this setting the one in which the highest degree of accountability is secured, of analysts to their ministers and chief officers and their disciplines, and of chief officers to the responsibilities of their post, of ministers to the boundaries that separate party from civil service, and so on.

The role of modelling and analysis in the federal Department of Energy, at least under George Bush senior's administration when the national energy strategy was being prepared (and still to a lesser extent today), exemplifies this rather well. Within the department, there is a great deal of teamwork and mutual trust between the rather long-serving group of senior policy and analysis staff, and much mutual respect based on deference to each other's competence in their respective disciplines of economics, law, engineering, environmental sciences, and so on. The priority for policy analysis is to combine intellectual credibility through synthesis with usefulness to politically appointed decision makers.

City mayor and public administration scholar John Nalbandian (1995, discussed in Frederickson and Smith, 2003, 30–1: see also the restatement of the argument in Nalbandian, 1999) emphasises exactly the present distinction between the nature of the information sought and deemed useful by elected politicians in leadership roles and by salaried staff analysts, when he defines the priority questions for each group as being, respectively, 'What do you hear?' and 'What do you know?', and the nature of their respective activities in the policy process as game-playing and problem-solving.

Although a policy analysis unit within a central executive agency may well be in a situation of great rivalry with those of other departments, it may – for example, in the case of a central body in the Cabinet Office or the office of the chief executive – have a certain authority over those of 'spending' departments. Even a group of units within a ministerial department, however, may well often be characterised by quite strong bonds because of their common membership of that department. There are good reasons to trust fellow unit members, to share information, to take part in relatively honest 'brainstorm' sessions together, to feel a sense of common identity, and to feel confident that one can trust colleagues with information obtained from elsewhere, for the institutional setting creates some common interests. Advancement and promotion are likely to be dependent at least as much on showing good teamwork in the production of analysis and the generation of ideas, as on individual intellectual brilliance. Individual hoarding of information is positively devalued (Feldman, 1989). A much more collegial atmosphere can reasonably be hoped for in such a setting. In short, in the professional policy unit, there are both high degrees of constraint from political and intellectual authorities, but also reasonable strong bonds at least internally, and some quite strong but authoritative rather than collegial ties can be expected vis-à-vis other parts of the salaried executive within which one has vertical links that are responsible for data collection, for implementation, and so on.

If backbenchers play snap! and if leadership candidates play poker, then the game played with information by the central units is best described as a form of chess: it is rigorous, well-structured, all the pieces are on the table for the others to see, there are marked differentiations in status between the pieces that can be played, and the range of acceptable moves are defined by the disciplines of professional policy analysis for a minister or a chief officer.

Football teams: the faction and the lobby group activists

Then consider the situation of the movement or the faction within a party. We might think of the Christian right within the Republican party in the US, or the Tribune Group within the parliamentary Labour Party, or the trade-union oriented left within Germany's Social Democratic Party, or the minority of free market liberals within the French Gaullist parties of the right. Faction members are united by a certain ideological commitment to certain sorts of principles, which they see the need to press upon a leadership of whose commitment to those principles they cannot be certain (Douglas and Mars, 2003).

Unlike leadership candidates and unlike staffers in professional policy units, faction activists have limited access to opportunities for influence, and few ways to influence how often those opportunities will arise. In this sense, they must act opportunistically, hoping reactively to mobilise disaffection with the style and the decisions of other actors in the policy system. This gives a certain episodic character to their political experience, for each opportunity assumes the character of a great struggle to build a narrative of confrontation between the principled stand of the faction and the pragmatic trimming of the leadership.

Within the faction, there are always dangers of schism (Lockwood, 1992; Douglas and Mars, 2003), for even those activists who aspire to lead factions tend to find themselves under pressure to trim, as they try to negotiate with and put pressure upon leaderships. It is hard for such leaders to discipline the faction, because while the bonds that bind faction members to each other through commitment to the mission are quite strong, it is difficult to legitimate the kinds of discipline of whipping and hierarchical subordination that characterises the relationship between the leadership contenders and the backbenchers generally or that describes the internal structure of the professional units. Factions produce hardline groups that are reluctant to delegate to their own would-be leaders, and more loosely attached moderates who cannot readily be disciplined.

While the internal bonds that unite the faction can be very strong, the obverse of this coin – and here their situation is similar to that of the professional policy teams within each ministry – is that they are strongly rivalrous vis-à-vis other factions. Strong membership cultures are also exclusive of those factions with rival principles. Alliances are therefore difficult, temporary and, to hardliners, potentially suspect for they run the risk of diluting the integrity of the principled message.

This situation gives the faction members what is, in theory, a strong accountability to the faction, but which in practice is much weaker than the faction can afford to acknowledge or imagine. The effort to hold leadership and bureaucracy to account is one waged through periodic confrontation, but is rarely fully successful in sustainable ways.

Factions tend to be generalists in their gathering of information, because their outlook upon policy is guided by principles for which they seek to embed a rather universal authority. They seek to apply their principles across a wide range of policy fields. Whereas for the leadership contenders, policy fields consist in opportunities for advancement and resources to be consolidated, and for professional central units, they consist in problems, and backbenchers have no consistent grounding in policy fields, by contrast for the faction, policy fields are merely sites upon which principled stands should be taken. The peculiarities of the problem structures of policy that preoccupy the professional analysts are of secondary importance for the faction: indeed, these characteristics are suspect as the kinds of excuses ('yes, but not in this case') that staff in these units wheel out for avoiding commitment to the kinds of principled action that factions demand.

Like backbenchers, factions find it hard to secure the kind of access to proprietary information about policy fields that is available to ministers and chief officers and their professional staffs. Therefore, they try to assemble bodies of semi-proprietary information about the nature of the normative commitments they want to make and see others make, and they build up banks of examples of how they can be applied, if only the leadership and the bureaucracy would develop the necessary political will. However, because of the limited trust within the faction and toward the other structures in the policy process, factions are often ambivalent about how far to impart such semi-proprietary information as they have been able to assemble. They fear that their work may be distorted, that it may be 'cherry picked' by leaderships and bureaucracies, thus ruining what they see as the integrity of their arguments, which, for them, ought to be seen and adopted as a seamless whole.

Factions devote time to sharing information among members, but they cannot afford their 'brainstorm' sessions to become too free, for the purpose of the internal discussions is to deepen commitment to and understanding of the integrity of the principles to which the faction is committed (Rayner, 1982, 1988). Information and knowledge in this setting are the basis of 'groupthink' (Janis, 1982).

Like the other forms analysed here, factions do not exist only among politicians. They can be found, most obviously, in the plethora of lobby groups, movements and pressure groups that have always crowded around the policy process, and they exist too in some niches of the salaried executive. For example, the associations of local authorities and health authorities can sometimes exhibit some of these characteristics of factions. Regulatory bodies which exist to promote principled action of various kinds can exhibit some of these problems: for example, the difficulties of managing schism and the uncertain legitimacy of leaders have long been problems in some of the equal opportunities regulatory agencies.

To represent the situation of the faction in relation to its members and to other actors as a game with information, it would be necessary to look at team games, such as football, in which transfers (defections) are possible, in which would be-leaders are distrusted and face the sack at unpredictably frequent points (club managers).

Mixed and unstable settings

There are a number of situations in the policy process that either exhibit mixtures of the characteristics of two or more of these four, or else that may by turns resemble any of them. For example, it was noted already that the situation of the congressional representative or senator in a system with a full separation of powers may be hybrid between individualistic and perhaps isolate forms. Nalbandian (1995) describes city managers in cities with that form of political constitution as often being a kind of hybrid between the game-oriented, insight-driven leadership politicians and the problem-oriented, expertise-driven staff analysts, precisely because their amphibious situation and Janus-faced accountabilities require this.

Fairfax County appears to exhibit features of a hybrid between several of these styles, each holding in check some of the excesses of the others. The supervisors and some of the highly entrepreneurial chief officers can operate quite individualistically and there is certainly competition between them for information and opportunities. However, there is little sign of the highly individualistic approach to praise and blame in decision making which is the distinctive feature of Baltimore's CitiStat programme and the culture that it is engendering in the city government. There remains a substantial element of institutionalised collegiality within the Fairfax system. Teamworking is the norm, including – albeit necessarily more attenuated by competition for resources – among chief officers. Staff are expected not simply to be specialists in a single discipline, but to develop collaborative working relations. There are certainly differences of grade, but none of the formality attached to the prized status of being able to 'work interagency' that is found at the DoE. The internally egalitarian tenor of the dress code, the common canteen and the office layout, designed for teamworking, the ability to use group decision support to sustain a more internally egalitarian style of deliberation, and the political structure organised around a chair of the board of supervisors rather than as dominant mayor are some of the symptoms of the presence of an enclaved form, but tempered by the presence of other styles of organisation. There seems, for example, to be little evidence of the problems of schism and dogmatism that can dog excessively factional organisations, which suggests that within that authority's informal institutions there are sufficient elements of individualism and hierarchy to constrain enclaving tendencies from becoming wholly dominant.

Select Committees of legislators in Parliament, or scrutiny committees of local authorities can take any of these forms or can combine them. Both of these are bodies charged with exercising oversight of the quality of policy making and implementation by the executive in a given area. There are cases where such bodies can work in ways that are rather similar to the central professional units, commissioning and working through extensive quantitative analysis in conventional forms, sharing information among their members, working in highly structured ways. Other such committees resemble pressure groups, focused on harrying the executive, looking for opportunities to blame. A third type of such a committee exhibits all the characteristics of the competitive cadre of executive leaders: members are rivals, many are more concerned to pursue individual careers in the executive and are only interested in using the scrutiny function as a stepping stone, and so they have no real incentives to share information with one another, and look mainly for opportunities to increase their own profile and only deepen their subject matter expertise where that suits their career strategy for breaking into the charmed circle of the competitive leadership. Finally, there are committees that resemble the isolate structure, not least because their backbench members are for the most part resigned to that role: they try to take advantage of such opportunities as come their way, but (rightly or wrongly) see little prospect of their being able to use the scrutiny function substantially to shape the agenda to which the executive must work.

Some wobble unevenly between each and all of these styles, but there are also scrutiny committees that combine features of just two or more of these styles quite stably over considerable periods. For example, the chairpersons of scrutiny committees may behave more like competitive leaders, while majorities of their members may operate more like faction members or isolates.

Consider next the audit bodies such as the National Audit Office and the Audit Commission (charged with examining public services to test for value for money) and some regulatory agencies such as the Office of the Information Commissioner (responsible for regulating data protection and, within the public sector, freedom of information) or the Parliamentary Commissioner for Standards or the Committee on Standards in Public Life (both with responsibilities for overseeing and promoting ethical standards among politicians). These regulatory agencies tend to combine, stably over the long term, the styles of the central professional units in their commitment to the disciplines of accountancy, economics and statistics, in their focus on a narrow range of variables that make up 'value for money' or legal propriety, and in their high degree of internal peer collaboration, with some of the characteristics of movements and factions, in their relentless and very public external pressure upon the executive to behave in ways that reflect the priority of the principle with the custody of which they are charged.

Situations and understandings of the policy process

Working in any of these situations leads people to develop particular lenses through which they are likely to see the policy process. More precisely, people tend to generalise from their own situation to the process as a whole.

In the chess players' setting, the policy process seems to possess a high degree of structure, and one can be confident that the information that is relevant to a policy decision is, from the unit member's point of view, neither unpredictable in the sense that it can spring from any number of sources (as for the backbencher) nor opportunistically garnered and dispensed wherever it is convenient in battles with other candidates for leadership (as for the political cadre or the permanent secretaries and chief officers), but structured by both the demands of one's political superior and the demands of the intellectual discipline of technical policy analysis. Moreover, the policy process actually does seem from this perspective to be structured into discrete projects, bills, programmes, evaluations, and cycles of 'stages' (as Lasswell, 1956 once conceived it: see Sabatier, 1991 and DeLeon, 1999 for continuing debates about the wider adequacy of this view on political science grounds) from analysis through to commitment, to implementation and appraisal through to re-analysis.

From the perspective of snap! players, the policy process appears to take the form posited by Kingdon (1995 [1984]). That is, the relationship between the events that create opportunities, the events that develop ideas and proposals and recommendations and the organisational capacities to make use of those opportunities and ideas, which appear relatively random. Therefore, it makes little sense to develop structured banks of information with which to try to make the most of such relationships. Random relationships of this kind are most likely to be generalised, beyond the backbenchers, in those areas where the prevailing institutional conditions push many other empirical groups into isolate positions. For example, on issues to which the executive devotes low priorities for analysis and modelling, where the main interest groups are weak and not worth the while of entrepreneurial politicians to appeal to, where media interest is spasmodic and elicited only by occasional scandals, and where either the nature of the policy field or the nature of the prevailing body of law produces a good deal of stability, then we can expect many of the actors to be effectively operating as isolates. For in such contexts, there are few reasons for people to form stable preferences, or seek extensive collective action or invest in knowledge management for policy, and every reason to engage in coping behaviours. It might be said that, for example, public policy around voluntary organisations exhibits many of these characteristics, and indeed, Kendall (2003) has argued recently that policy makers in this area have behaved in ways that render the Kingdon model usable.

For the individualistic poker players, whether elected or appointed, the policy process appears to have the structure posited by theories stressing the role of policy entrepreneurs in networks of other individuals, much as presented by Laumann and Knoke (1987) and Knoke (1990). Here, rhetoric, persuasion and selective gathering and deployment of proprietary information are seen as central, as Majone (1989) argued. Again, there are policy fields which genuinely do seem to exhibit characteristics of this kind. There are areas of foreign policy, especially in those areas which are least subject to multilateral institutional governance and left largely to the discretion of states, in neo-realist fashion, in which diplomacy readily and understandably becomes a poker game, and in which actors assume that everyone else plays by the rules of poker. Allison and Zelikow (1999 [1971]) show that, seen through some of their theoretical lenses, and especially from the individual points of view of Kennedy and Khrushchev, the decision-making process during Cuban missile crisis of 1962 appeared to have this structure. For although each was constrained to some degree externally and by each other's actions, they were both able to manoeuvre with comparatively significant freedom internally and with respect to each other and between their own civilian and military leaderships.

From the point of view of the faction members, the policy process looks like a battle of competing factions or advocacy coalitions, much as Sabatier and Jenkins-Smith (1993, 1999) portray it. Again, there are policy fields that seem to push many of the actors into enclaved situations. It is not surprising that many of the studies on which Sabatier and Jenkins-Smith (1993) developed their model were ones – such as industrial policy and environmental policy – which are structured around highly mobilised and ideological interest groups and where laws are often overturned when the balance of power shifts.

Institutional styles

A taxonomy of information settings for policy making

The four settings presented so far, are of course, archetypes or ideal types, even stereotypes. However, they enable us to appreciate something much more important than any old bunch of stereotypes would do, about the conditions in which different kinds of e-governance tools are actually used and found to be useable. For they have been distinguished from one another by two key but quite general dimensions which are the most important classifying factors for the setting of policy makers in the policy process, namely the degree of *constraint* and the strength of the *bonds* to (in the first instance) peers and (secondarily) to other actors in the policy process. These two dimensions are, a century of social science has shown, crucial to understanding the conditions under which action, motivation, information gathering and use, understanding and knowledge management, are achieved.

These are Durkheim's (1951 [1897a]) concepts of social regulation and social integration, or (1961 [1925a]) discipline and attachment, which were his key measures of social organisation. What Durkheim should have done in that work was to cross-tabulate them, rather than treat them as having little to do with each other (Thompson *et al.*, 1990, 138). This was first done by Douglas (1970), who termed the two dimensions 'grid' and 'group'. Since then, the approach has yielded a huge body of research showing the ways in which the two dimensions structure the basic variety of situations in which information is used, ideas are developed, decisions are made, and so on. When cross-tabulated to yield strong and weak segments, the two dimensions identify the four basic institutional styles which shape political, social and economic life in any system of social organisation, at all but the very smallest scale: (see e.g., Douglas, 1978, 1982a, 1992a, 1996; Gross and Rayner, 1985; Schwarz and Thompson, 1990; Hargreaves-Heap and Ross, 1992; Coyle and Ellis, 1993; Adams, 1995; Ellis and Thompson, 1997; Hood, 1998; Mars, 1982; Thompson *et al.*, 1999; 6, 1998, 1999a,b,c, 2001a,b, 2002a, 2003).

The matrix identifies clearly four basic *institutional styles*[8] of organisation and relationships among policy makers, set out in Figure 8.1. These styles can be described in general as isolates, hierarchies, individualism and enclaves, of which the cases discussed in the previous section provide clear examples.[9]

Apart from the pure cases with which we have so far been concerned, there are – as we have seen – mixed cases, and there are also 'settlements' between two, three or four of these institutional styles in which compromises are – at least tacitly and implicitly – struck.

These are not psychological types. No one is born a factionalist or a competitive cadre member, nor remains one having once been in these

Constraint ⇑	Loose clusters of isolates	Central units, hierarchies
	Strong constraints, weak bonds	**Strong constraints, strong bonds**
	e.g. *snap! players*, backbenchers; some weakly coordinated administrations	e.g. *chess players*, disciplined ministries, senior policy staffs, professional policy analysis units, highly cohesive cadres
	Competitive individualists	**Enclaves, sects**
	Weak constraints, weak bonds	**Weak constraints, strong bonds**
	e.g. *poker players*, leadership contenders, rival ministers, power brokers, political entrepreneurs	e.g. *football players*, factions, some party groups, some collegial boards

⇒ **Bonds**

Figure 8.1   Four basic institutional styles of organisation of policy makers

situations but having moved to another. Rather, these tendencies and capabilities are the product of the basic institutional situations in which policy makers find themselves.

In this chapter, I shall argue that these different strengths of forces of constraint and bonding, when in the presence of the other, have important causal effects,[10] that shape what information policy makers in each of these situations want, what they can use, how they will frame the information they seek and obtain, how they will make sense of their problems and experience and how they will make decisions. For once we know what kinds of information people in different institutional contexts will be led to seek out, we can begin to understand what style of political judgment they might seek to cultivate, and, in the final stage of the argument, how they might approach the choice and the use of e-governance tools to support them in these activities.

Institutional styles and information

We can now summarise the analysis presented above in a series of versions of matrix as shown in Figure 8.1, to explore how the four basic styles work to structure the conditions under which e-governance will be pursued. First, consider the key issue for our present purposes, of how these institutional styles of organisation and relationships induce policy makers to gather, use and trust others with information. For, this once understood, it becomes possible to understand how they can use e-governance tools in the course of policy making. Figure 8.2 summarises the findings of the analysis given above.

In this figure, it is important to attend to the similarities between the styles vertically, horizontally and diagonally. Consider, for example, the diagonal connections. For the diagonal between the hierarchical and the individualist styles represents the most active and specifically proactive stance toward information and the one that is most oriented to understanding policy making as the making of specific decisions. By contrast, the diagonal between the isolates and the enclaves is the more reactive, more oriented to policy making as a challenge and a danger.

These alternative styles of managing information and knowledge are, as we shall see below, the fundamental determinants of what policy makers in different settings can actually do with e-governance tools, and also of the particular tools they will find most congenial. However, before we can appreciate fully the impact of the institutional organisation of policy making upon the use of e-governance, it is important to understand the differential nature of learning in general – and therefore

Constraint ⇑	Loose clusters of isolates	Central units, hierarchies
	Generalism of an eclectic kind	Specialism, but principally for sharing within teams
	Access to information mainly limited to what is in the public domain	Access to proprietary information
	Only occasional and unpredictable access to influence	Routine access to power and influence
	Information and knowledge as casual resource	Information and knowledge as analytical tool in procedures
	No reason not to trust peers with information but few incentives or occasions to do so	Imperative to trust peers with information, to share ideas and develop analyses and recommendations jointly
	Fragmented experience	
	Policy fields need have no particularly consistent character	Structured experience
	Information game analogy: snap!	Policy fields consist in problems and solution designs
	There are no problems, properly speaking, just conditions that have to be coped with; sometimes chronic conditions turn into acute crises ('Events, dear boy, events', Harold MacMillan)	Information game analogy: chess
		Problems can be sorted into the soluble, the manageable and the insoluble but containable: crises are evidence of failure to classify and manage or contain problems

⇒ **Bonds**

	Competitive individualists	Enclaves, sects
	Specialism, but principally to secure proprietary control over information	Generalism, focus on universal principles
	Access to proprietary information	Difficulties in securing access to proprietary information; efforts to create some private information within the enclave
	Continuous access to power as long as position is maintained	
	Information and knowledge as ammunition	Periodic, unpredictable access to influence
	Incentives to distrust, to keep information private from peers	Information and knowledge as groupthink
	Disjointed experience	Difficulties in securing internal trust, reasons for external distrust
	Policy fields are opportunities and resources	Episodic, shared experience
	Information game analogy: poker	Policy fields consist in sites for principled action
	Problems and crises are opportunities	Information game analogy: football
		Problems indicate that principles have not been followed; crises are occasions for confrontation over principles

Figure 8.2   How the institutional styles gather, use and share information

of any expectations for and commitments to computer assisted learning – in these different settings.

Institutional styles and learning

Digital tools to support policy makers are usually presented by those who design them and those who write manuals as tools to support 'learning' by policy makers. At a time when the business and management communities are peculiarly interested in how to induce organisational learning, especially among board members and senior management (e.g. Senge, 1990), this is hardly surprising. However, talk of 'learning' in public policy-making contexts can be a highly fraught affair, and it can neither be assumed that policy makers in all situations welcome learning nor that what is learned is necessarily more accurate, relevant or useful than what is unlearned nor indeed that policy learning is essentially the acquisition of more or different technical information about problems and possible solutions. Politicians can often feel that there are costs, not only in being seen by voters as standing in need of additional learning, but also in the stability of their authority over their parties and, when in office, over the bureaucracy, as well as costs in weakening their negotiating position with lobbies and pressure groups. At election times, it has long been thought risky to admit to be learning, for politicians in many countries typically believe that to admit to learning is a confession of inadequacy. For the advocates of the 'learning organisation', this may seem to be reminiscent of the old-fashioned authoritarian management style that they want to challenge in business, and something that politicians simply need to move beyond. Likewise, among political scientists concerned with understanding the policy process, there are some who write about 'policy learning' as though it were essentially a matter of the changing of beliefs to more accurate ones, the adaptation of preferences to more feasible policies, or of making more effort to take account of the findings of evaluative studies on programmes (e.g. Leeuw *et al.*, 1994). However, there are important and very real institutional constraints on what policy makers can understand by 'learning', and these are not confined to the importance of ideological commitments as limiting what is acceptable to be learned (contrary to the view argued by Sabatier and his collaborators: Sabatier and Jenkins-Smith, 1993; Sabatier, 1999: for a critique of policy learning as simply the acquisition of technical information and of the narrow role of ideology in Sabatier's work, see Fischer, 2003, ch. 5).

Using the analysis of constraint and bond as explaining the conditions for learning, we can see that learning will be both practised and

understood differently. Here, we focus on learning for the improvement
of each policy maker's capabilities to perform the roles available to her
in the particular institutional setting she finds herself as perceived by
the policy maker and those others in her or her own institutional setting, rather than (say) learning that might be relevant to roles she might
hope to take in the future or simple acquisition of technical information
about policy problems and instruments for its own sake. Figure 8.3
applies the argument developed in the discussion above to this problem.

The case studies provide examples of this. For the individualist Mayor
O'Malley, key pieces of learning are the scraps of information by which
he can detect poor performance: hence, he is careful to make full use of

Constraint ⇑	Loose clusters of isolates	Central units, hierarchies
	Learning is hardly possible, because experience is fragmented and random, and what is useful today may be of little use tomorrow	Learning is important, because experience is structured and continuous: the past holds lessons for the future
		Learning is the acquisition and deepening of formal skills of policy analysis and of oversight and management of policy execution
		A little knowledge may be more dangerous than none: the boundaries of the decision and learning group should be set, and skills and factual understanding should be graded according to status
		Policy makers can afford to admit publicly to needing to learn on the job: it can even enhance credibility, provided that it is done sparingly.
	Competitive individualists	Enclaves, sects
	Learning is mainly noticing of scraps of disjointed information, that might indicate opportunities	Learning is the collective process of raising consciousness, of deepening commitment to principled action.
	By the time one is a decision maker, it is too late for learning basic skills: you either have them or you don't	
	Never admit publicly to needing to learn.	

⇒ **Bonds**

Figure 8.3 How the different institutional styles of organisation understand
differently, learning for improvement in their capabilities in the policy process

his car journeys around the city and of the CitiStat analysts' tours with their cameras. However, his expectation is that his agency chiefs will not learn simply to correct these particular cases, but will respond by re-examining their systems as a whole, and engaging in continuous system-wider learning, in a more hierarchical fashion – the fashion that it is neither possible nor appropriate for him to use in his own mayoral role. Learning in the more collegial world of Fairfax County can sometimes seem mildly enclaved because of the presence of the other forms of organisation: for example, having become committed to the use of GIS and to the importance of geographical analysis as a principle of local public service management, it is hardly surprising that what gets learned tends to reinforce the importance of this style of analysis and policy development.

Perhaps more importantly. Figure 8.3 shows that policy makers in each institutional style of organisation must specifically reject the kinds of learning that the other styles are committed to, as being at least (in the case of the isolated backbencher) pointless, or inappropriate or even positively dangerous to their commitments. Policy judgment is then a problem of appreciation of what counts as relevant, a problem of classifying types of information – essentially a problem of achieving more rather than less intelligent *information-rejection* (Thompson and Wildavsky, 1986). Policy makers, like everyone else, need to reject huge quantities of information as less relevant, as less important, as impossible for them to process or take seriously without compromising fatally their ability to function, or they could not allocate the scarcest resource in government, namely, attention to problems (March and Olsen, 1975; Simon, 1977).

The ambitious minister sees dangers of dogmatism in the kind of learning that the faction demands, just as the faction activist sees in the ruthlessly pragmatic learning of the minister the danger of undermining moral commitment. The minister is happy to take the learning of the professional unit as grist for her political mill, but her real concern is not so much the technical quality and long-term efficacy of policy design as knitting together a story and a personal strategy with short-term electoral success for the party. Because the individual backbencher can do little to influence even short-term electoral prospects, save at the margin in her own constituency through assiduous constituency work, most of the kinds of learning that might be valued by the others is of limited value to her.

As we shall see shortly, these commitments and rejections about learning make significant differences to the ways in which policy makers in different settings think about the kinds of activities that particular e-governance tools might be of some help with.

Institutional styles and political judgment

Throughout this book, I have argued that the principal reason for being interested in e-governance and the particular tools and software packages that support it, is to explore what contribution it can make to political judgment. Therefore, we need to understand the different ways in which policy makers in these different basic institutionalised settings exercise political judgment differently. In Chapter 7, I argued that political judgment should be understood in the general case as comprising five elements. Now we can see that the ways in which judgment is shaped, given its specific hue by the situation of policy makers can best be understood by exploring the ways in which institutional styles will stress different dimensions from among those five.

In the situation of the backbencher, whose opportunities to originate either the presentation of policy problems or of solutions are few, and for whom policy problems in her constituency represent threats to her re-election prospects more often than they represent opportunities, judgment is principally a reactive affair. She has no particular reason to stress any of the five elements in judgment more than any other, for she is not in control of where she must allocate attention; her information sources are restricted; her opportunities to influence the framing of problems are limited by pressures from constituency lobbies and from party whips; and she has few opportunities to influence the process of closure. Sense-making is for her, a hand-to-mouth affair, for the principal activities of sense making are conducted by others. She is not so much a 'policy taker' – for backbenchers play an important and in any case, an irremovable role – as a sense-taker. Judgment consists in finding a way to *cope*: coping strategies for different crises are often no more integrated than are the bonds between these comparatively isolated policy makers.

By contrast, for the activist in the party faction or in the agency dedicated to the institutionalisation in the administration of a moral commitment, the key dimension of judgment is the third, or process-oriented one. For the most important thing in political judgment from the perspective of the lobby group or faction is to situate each issue, problem or crisis in relation to its preferred principles and commitments, to make sense of it within the moral narrative around which the movement is organised. In this context, however, the allocation of responsibility and blame for failure to follow principle is a key part of the faction's role in judgment. Judgment consists in the *subsumption* of problems under universal moral concepts: information is therefore integrated just as the bonds of accountability in the faction are.

For the professional policy analysis unit, the central focus of time and effort, and the priority is always given to the second and fourth element of judgment, namely, the backward-looking analysis of causation and the forward-looking process of the generation and appraisal of options. For the professional policy analyst, the larger narrative can be taken for granted: if it is an economic analysis team, then the larger picture is of course the economy and the problem is framed as part of a narrative of economic governance, and similarly for legal, social statistics, and other disciplinary styles of policy work. Unlike the minister or chief executive, the policy analyst has limited choice about where to allocate her attention: the priorities are settled by politicians and managers. At most, she can lobby about where the first element of judgment should be directed. Finally, the fifth element, of taking responsibility, is also outside her remit: it is for politicians and senior managers to shoulder whatever trade-offs she identifies and to accept as a price worth paying or else as a resigning matter, any unintended consequences that the professional analyst has not identified in advance. Judgment for the professional is an analytical, cognitive and *inferential* process, amenable to codification: the structure of information in judgment once again recapitulates the structure of the bonds between the individuals in the disciplined agency, being both regulated and integrated.

Consider finally the position of the minister, the permanent secretary, the agency chief, or the chief officer. For her, the first dimension of political judgment is the biggest challenge, and the riskiest if got wrong. However, it is the element of the judgment process that she can neither duck nor delegate. Of the backward-looking dimension, the main focus of interest is the allocation of blame and responsibility. The third dimension is her treasured role, namely, that of fitting political experience into the larger narrative, which is her key contribution to sense-making. Finally, the fifth element is her peculiar burden: she must shoulder responsibility for the consequences of decisions, welcome and painful or embarrassing alike. For the development of options, she depends upon the professionals: for the development of alternative narratives, she must look to the lobbies and the factions. For the leadership candidate, then, judgment must be conceived as a process of *insight*, in which she must very rapidly soak up material from professional analysts, movements and isolates, work with a rough and ready appreciation of priorities and follow her overall sense of the general narrative to which she works. Inference and analysis are secondary to appreciation, principle to pragmatic balancing and distancing, all of these secondary to closure and responsibility. In judgment as insight, again, there is little integration or regulation in the ordering, acquisition or use of information just as these features are limited in the bonds of accountability than the individualistic cadre members face.

Figure 8.4 summarises the argument and cites some paradigmatic texts of social science or philosophy that exhibit accounts of decision making in each of the styles.

There are some important diagonal contrasts to be noted here. To exercise judgment, in policy making and politics, is often understood as being implicitly doing so from a situation of power, of being able to secure continuous influence over what gets done and how it gets done. We have noted that the situations in which this experience is most likely to be found are those of the leadership contenders and the central professional policy units. Comparing their conceptions of what they do,

Constraint ⇑

Loose clusters of isolates

Judgment is *groping for coping responses* (e.g. after events experienced as shocks) (Wildavsky, 1988)

Immediately oriented judgment is key: find a way to get through the day's urgent crises

Central units, hierarchies

Judgment is *analytic inference*: methodical problem structuring, prioritisation of competing demands for attention, explicit goal-setting, explicit instrument selection, under conditions of bounded but real rationality (Simon, 1997 [1945])

Process-oriented judgment is key: develop an adequate and rigorous understanding of the structure of the problem

⇒ Bonds

Competitive individualists

Judgment is *insight*: a non-inferential craft skill of recognition and appreciation of the contours of problems, resources, opportunities, capabilities, possibilities (Barnard, 1938b; Steinberger, 1993)

Immediate, forward and backward looking judgment are key roles: allocating attention, allocating blame, both devolving and shouldering responsibility

Enclaves, sects

Judgment is *application of principle*: identification of appropriate universal concepts and principles, generation of consensus around them, commitment of political will to apply the principle (Kant, 1987). Backward-oriented judgment is first priority: identify responsibility for problems that arise

Figure 8.4 How the different institutional styles of organisation conceive of political judgment

when they exercise judgment is therefore instructive. For there are essentially two ways of defending judgment when one is in a position of power: either one can claim that is an essentially unanalysable craft skill that can be acquired but not taught, as for example cultural élites have long done in respect of their aesthetic judgments (Bourdieu, 1984 [1979]; see also Douglas, 1981, 126), or else one can do as priesthoods have always done, and argue that judgment is indeed analysable, because it consists in the conduct of analysis, and that it is teachable but requires exceptional rigour, knowledge and intellectual capability that only those who have been extensively and properly trained can expect to achieve. These are, in fact, the contrasting strategies of the leadership and professionals on the positive or 'power diagonal'. By contrast, the claims to judgment as coping or as principle express the aspirations of the less powerful to discipline those who do exercise power to address their material or their moral concerns, as they arise from their peculiar situations. For one can challenge the judgment of the powerful either on the grounds of its haughty irrelevance to actual fate, or on the grounds of its iniquitous disregard for the moral claims of principle that make up some strand of the moral culture of a society: one is resigned and the other radical. These are, indeed, the conceptions that emerge on the other diagonal that runs across the institutional styles of power, from survival to resistance.

This chapter has argued for a theory which might be called a political science of policy knowledge, by analogy with the sociology of knowledge. The central insight of the century-long tradition of the sociology of knowledge is that the manner in which people develop ideas, lay categories and classifications and use information, essentially *reproduces in those categories and in the structure of the information they seek to build or use in judgment, the institutional features of their own social organisation and accountabilities*. This is in effect a restatement in more contemporary terms of Durkheim and Mauss's (1963 [1903], esp. 11) argument that the forms of classification people use for features of the world in general are derived from the forms of classification they understand from their own social organisation. Whereas Durkheim and Mauss tried to argue for exact mapping and transposition of substantive elements, the present argument is that it is more fruitful to recast the claim as being at the level of the structure of relationships, and in the larger *thought style* (cf. Merton, 1945; Fleck, 1979 [1935]; Douglas, 1986; Durkheim, 1995 [1912a]; Zerubavel, 1997). The present argument is that the institutional features of the *micro-political* organisation in which people are differently situated in the policy process are reproduced in their manner of seeking, organising and using information in exercising political judgment. For example, the politically and socially strongly integrated community of hierarchy is reproduced in the highly structured and integrated sets of explicit data to which people appeal, in their policy work in these settings. By contrast, the weak bonds to others that are the mark of the backbench or stranded bureaucrat as isolate are reproduced in the weakly integrated and informally linked data which they must use to survive. The bodies of deliberation upon principle, which are the great libraries of the enclaves, are strongly integrated in content and literary style but weakly structured by comparison with the formal data sets required for modelling or simulation. The individualistic leader's preference for unstructured proprietary information, from which she can borrow but to which she need not contribute, reflects exactly the social situation of such people, as weakly regulated enough to behave entrepreneurially to secure control of the information that matters to them, and weakly integrated enough for them to hope to prevent unauthorised others in many cases from gaining access to their data. The styles of judgment and the selection and use of judgment technologies follows clearly from this mirroring of social and informational structure.

With this account of the institutional constraints upon collaboration and rivalry, and of the information games that differently situated policy makers are required to engage in, it becomes possible to explore how different kinds of e-governance tools will be selected and used in each institutional style, and in situations characterised by settlements between them. This will be the focus of Chapter 9.

9
Institutional Styles and E-governance Tools

If the argument of the previous chapter is accepted, then it should come as no surprise that these different institutional styles will look to different kinds of software with which to provide themselves with support for their peculiar contribution to the process of governance and judgment, worry about differently and have different aspirations for each of the principal kinds of software available to them.

This chapter will explain why different e-governance tools are adopted by policy makers in each of the institutional settings, before examining the different styles of exercising judgment, making decisions, and embedding these e-governance tools in ritual-like processes of judgment, before concluding with a discussion of the scope for using e-governance tools in each setting to sustain learning among policy makers. Returning to the theme identified in Chapters 2 and 3 that the procedures, the disciplines and the physical setting of policy work matter enormously, the chapter concludes with the argument that we need to see the use of e-governance tools in their ritual setting, for judgment is rightly and properly a highly ritualised process. Contrary to the crass view that rituals are somehow insincere, trivial, or deliberate exercises in mystification, the exercise of ritual is in fact a sign that something is considered worth taking with great seriousness and commitment, and important enough to be grounded in some of the most fundamental practices available to human beings. The argument is that in each of the four settings, different ritual practices of judgment are adopted, and therefore the use of e-governance tools must serve these ends.

Choosing tools

Following their role in political judgment as providers of information, analysis and inference, and their commitment to learning as the deepening of understanding of policy issues as problems and the appraisal of technical solutions against definable standards, it is the central professional units where the greatest interest is displayed in the more advanced tools of e-governance. Thus, it is hardly surprising that the DoE is more interested in modelling than either Fairfax or Baltimore. It is not only that modelling is expensive, and demands large budgets, although this does limit the demand for it from local governments: for there are sub-national authorities using modelling – in Chapter 3, Rohrbaugh's (2000) modelling work with New York State was summarised. Moreover, not all modelling is prohibitively expensive.

Also important is the nature of the organisation. Because sharing of information and a high degree of both trust in peers (bonds) and disciplined styles of analysis and disciplined accountability for priorities (constraints) characterise their situation, it is possible for these units to organise their use of these tools explicitly around learning within the context of their highly structured experience of the policy process. In these settings, highly structured groupware can be used to its greatest advantage. Structured databases can be assembled that record not just contacts made and messages exchanged, but programmes of joint argumentation and debate, bodies of both quantitative data acquired and made available for re-analysis and of hierarchically organised banks of documents organising ideas, problems and analyses conducted internally or borrowed from other sources with similar capabilities and analytical cultures across government or from academic institutes, consulting houses or think tanks. Knowledge management – in more than one of the senses in which that umbrella term is used (see Chapter 3) – can be conducted within such units in quite structured and powerful ways. For example, in these settings, it is possible to make good use of highly structured procedures of decision support, such as decision conferencing tools, meeting management software, and to subsume these uses within overarching environment tools such as intranets and the use of project management software to govern the organisation of policy analysis in a structured series of projects.

From the position of the backbench isolate, these tools offer rather limited purchase on the problems she faces. Because she has only randomly occurring opportunities for influence over policy decisions, and because she has limited and randomly occurring access to the kinds of proprietary and structured information that is the fuel of the central professional units, the advanced tools of e-governance seem to be pretty toys, but at once over-powered to be worth investing in, given what she can feasibly achieve. Yet they may well seem curiously abstracted from the grind of day-to-day miseries that she encounters in her 'surgeries' in which constituents can come to see her for help with particular problems they may have in their dealings with public authorities. For the most sensible digital tools are those that support the kinds of quick response to influence opportunities, making occasional responses to those initiatives from the professional units and the political or administrative leadership that her special interests and other pressures lead her to make priorities, but only for the time being, before other pressures shift her attention elsewhere in ways largely beyond her control. For her, then, individual e-mail accounts, extensive use of word processing, provide ways to provide inputs to central collections of ideas, knowledge and information about conditions, problems and preferences. At most, the local representative may be interested in some kind of customer relationship management (CRM) tool with which to manage her or his communications with constituents. It is interesting to note that this has been one of the key demands of the Fairfax County elected supervisors for support: they have – as the theory would predict – shown much less interest in, for example, meeting management tools or direct access to county government datasets than do the senior executive staff. Use of new information technologies is often undertaken by isolates on a half-hearted and occasional basis, by contrast with the disciplined way in which inputs are expected in the name of knowledge management and organisational memory capture, within the central professional units.

Communication and data handling have a quite different kind of importance for factions and movements, but again this importance flows from their conception of political judgment and of learning, and in turn from their situation. Like the position of the central professional unit, within the movement or faction the sharing of information is essential to the task in hand. Unlike the central professional unit, however, achieving trust between members that each is indeed sharing fully and honestly is much more difficult, because of the fear that such limited quantities of proprietary information that are assembled within the enclave may be monopolised by a leadership that cannot be fully trusted by the members not to compromise too far with the other players in the policy process. In order to sustain itself, the movement or the faction must support and deepen its members' understanding of the general principles and commitments that define the purpose and meaning of

the movement, its *raison d'être* and its demands, for this is what judgment consists in and this is the only real meaning of learning in this
position. Therefore, relatively unstructured, un-moderated, leaderless
forms of information sharing are highly prized, and in particular, information tools that enable the expressive culture of passionate commitment to principle to be expressed – and the deviations of others to be
detected and sanctioned – are of particular importance. The Usenet
world or the e-mail listerv is therefore a peculiarly appropriate form of
communication, but also, from the movement's perspective, the Usenet
or listserv archive is an important form of knowledge management.
Argumentation support tools such as electronic conferencing systems;
more generally, discussion and debate packages that do not force people
into inappropriately early closure or coarse forms of procedure such as
simple majority voting that either risk schism or inattention to principle, are particularly valued. Again, as the theory would predict, it is the
more internally egalitarian Fairfax which can make good use of a group
decision support centre designed for collaborative meeting management. Fairfax supervisors are regularly e-mailed about developments,
although they still want agenda papers for meetings to be sent to them
in printed form. Most such communications come as much from the
executive as from the chair, unlike mayorally dominated Baltimore city.
In enclaves, e-mail circulars and listserv systems are useful, provided
that they are not seen as ways in which an insecurely grounded leadership can impose its position and potentially excessive pragmatism,
upon the core and committed membership.

Finally, consider the situation of the competitive leadership members
and leadership candidates, both elected and administrative. Their most
important information requirements are for ideas, for ammunition in
defence of commitments, for proprietary control over information that
would advantage their rivals in, for example, budget setting, and for
information in advance about crises and opportunities, which also
should ideally not be available to their peers. Like all managers, they get
the vast bulk of their information from face-to-face contact, and when
drawn from written sources, they rely much more on documents than
on quantitative data sets (McNurlin and Sprague, 1998 [1993], 215).
Mayor O'Malley, for example, is said to have time to do no more than
skim read printed memos from the CitiStat programme and to pick up
much more information in the review meetings: similarly, the political
appointees in the DoE read printed reports with a handful of simple
charts; they have neither time nor need to consult the statistical data
directly. It is not merely that such people are able to delegate what they

find to be the drudgery of quantitative analysis to the central professional units, although that is typically also true. Rather, the reason is the more important one that what they find really valuable is information that provides interpretation of experience, rather than formal data: those data are of course already under some implicit interpretation but they still await more coherent and action-oriented interpretation at the high level in order to enable cadre policy makers to use the information to identify opportunities and engage in strategy development. Their peculiar skills in learning and judgment are in lateral rather than linear thinking: as we have seen, they look for connections between disjointed pieces of information that provide clues and hints of the ways in which trends are moving, and they see to control who can rival them in access to these interpretations. It is therefore hardly surprising that those e-governance tools using groupware which only work effectively when participants can all be assumed reliably to be sharing information candidly, are not ones to which leadership contenders, ministers, agency heads and departmental chief officers resort to readily. Indeed, the fact that work has been done to produce large banks of shared data is enormously valuable to them: put rather harshly, that is an essential public good for all leadership members and candidates on which each member or candidate is a 'free rider'. They have few reasons to be assiduous or disciplined in their input of such information, but every reason to draw opportunistically and ruthlessly upon the information in banks accessible over intranets, that the central professional units have assembled, especially from their own proprietary information stores, but also from consultations with isolate backbenchers and movements and factions of every kind. For their communications, secure e-mail is vital to their capability to build their temporary and unstable coalitions with one another in order to secure votes in leadership challenges, to secure advantage in budget negotiations, to ensure support when they are in difficulties with the public media or with superior administrative or political leaders. The privacy of their control over their storage of their own word processed documents is a paramount concern: unlike backbenchers, careful storage is a priority, for they may need later on to be able to defend themselves against misinterpretation of their own documents, however limited their circulation.

Figure 9.1 summarises the argument of this section about the kinds of e-governance tools to which people in each of these situations will most readily make resort.

However, these considerations about how the different situation of players in the policy process shape their willingness and indeed their

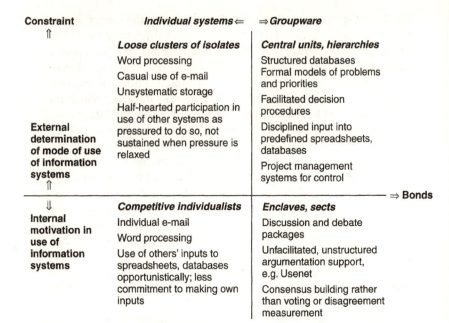

Constraint ⇑	*Individual systems* ⇐	⇒ *Groupware*
	Loose clusters of isolates	***Central units, hierarchies***
External determination of mode of use of information systems ⇑	Word processing	Structured databases
	Casual use of e-mail	Formal models of problems and priorities
	Unsystematic storage	Facilitated decision procedures
	Half-hearted participation in use of other systems as pressured to do so, not sustained when pressure is relaxed	Disciplined input into predefined spreadsheets, databases
		Project management systems for control
		⇒ **Bonds**
⇓ **Internal motivation in use of information systems**	***Competitive individualists***	***Enclaves, sects***
	Individual e-mail	Discussion and debate packages
	Word processing	Unfacilitated, unstructured argumentation support, e.g. Usenet
	Use of others' inputs to spreadsheets, databases opportunistically; less commitment to making own inputs	Consensus building rather than voting or disagreement measurement

Figure 9.1 How the different institutional styles of organisation look for different kinds of software in supporting their roles in governance, and how they make different uses of the same pieces of software

ability to use different tools of e-governance do not exhaust the shaping effects of situation upon e-governance. These basic institutional styles are of central importance in giving structure to battles over what kinds of e-governance should be used *by other* players in the policy game, and what they should be used for. For the really important role of these institutionalised situations is not simply in shaping what people are themselves prepared to do, but in what they will resist others attempting to do – that is, in the learning they will reject. In order to understand this, it is helpful to think about the doubts, reservations, concerns and fears that people in these different situations will have of the ways in which – and, more importantly still, the purpose for which – others in different situations may make use of the tools of e-governance.

We may begin with the position of the backbenchers. Sandwiched between pressures from whips, their constituencies, the latest issues picked up by the media and the apparent fickleness of their local and national electorates, backbenchers worry that e-governance tools are likely to be used to distract the attention of both the leadership and the professional policy administrators from the immediate pressures that worry them most. The fact that professional policy analysts focusing attention upon long-range strategy, seems to such backbenchers and other comparative isolates to take their eye away from what seem to them the really important short-term issues of practical and quite confessedly short-term problem solving. The language of 'delivery' and 'visible action' comes most readily to this group. Moreover, their situation makes them peculiarly attuned to the phenomena that represent the deformations and limitations of rational policy analysis, and the raw cussedness of the events. The very randomness of their experience of policy and politics attunes them to the ways in which rational action can exacerbate randomness and noise in policy. Goodhart's law, formulated by former Bank of England and now academic monetary policy economist Charles, Lord Goodhart, states that any variable that is targeted by policy makers because it seems to have some coherent and steady relationship with some outcome that policy makers care about – for example, one or other measure of the money supply and inflation two years hence – will lose that relationship over time precisely because it is the target of policy action. In short, the effort of policy makers to target an indicator in programmes will itself cause that measure over time to cease to be a good indicator of the underlying problem it was meant to capture (Jackson, 1992). A special case of the laws formulated by those stout analysts of public policy, Sod ('if it can go wrong, it will') and Murphy ('Sod was too optimistic: it already has gone pear-shaped'), Goodhart's law reflects the wisdom and experience of the loose clusters of isolates everywhere in the policy process, and not only on the backbenches. From this, there follows a generalised anxiety about the ways in which the professional bureaucracy tends to invest in rational policy analysis, in models that seek to anticipate the future, when the very use of those models and the very effort put into rational policy making simply exacerbates the universal tendency towards unanticipated consequences (Merton, 1936). Similarly, the free-wheeling games and the hoarding of information by ministers, leadership contenders and the most senior bureaucrats seems to the isolates unlikely by some invisible hand to lead to greater success in securing re-election or in controlling unanticipated consequences, while the endless discussions of principle among the movements and factions seems to exhibit little understanding and still less willingness to gather intelligence about the day-to-day problems experienced by their constituency, the fickleness of the public toward parties, and the sheer unprincipled pragmatism of the voters.

From the perspective of the leadership contenders and the departmental chief officers, the prospect of conducting public policy through endless philosophical debates on line about principle and how to apply it to particular problems, seems ridiculous, lacking any grip on the changeability of problems and events. From their point of view, a focus on grand principles of justice shows the same tin ear for real politics as does an obsession with mathematical modelling. For they are also fearful of the agenda of policy rationalism implicit in the use of e-governance tools by the advocates of the professionalisation of policy making. What they understand by judgment is precisely the overriding of technically sound recommendations based on empirical modelling of policy problems, where political realities and opportunities seem to their own insight to demand it. Therefore, the idea that they should themselves submit to the rigid disciplines of professional policy design threatens deeply their understanding of political policy making as a craft of interpretation, in which watching the signals that come from face-to-face communication and debate is at least as important as, if not more so than, sound economic, environmental or social policy analysis. For example, a recent survey of studies in the health field found that senior policy makers took research evidence more seriously if they had personal contact with the researchers, a finding consistent with this view of the prevailing institutions of individualism (Innvaer *et al.*, 2002). Mayor O'Malley was attracted to CitiStat precisely because of the centrality of the face-to-face ritual of cross-examination, and because it represented a way to secure political control over a bureaucracy which he saw as failing, and a way to open up the closed box of departmental loyalty and certainly not because he wanted decision making in any simple way to follow from the numbers generated. The professional use of e-governance tools threatens to bring closure to the policy networks that individualistic leaders strive constantly to keep open, palpable and fluid. Just as the isolates know one big thing about unanticipated consequences, so leadership contenders and senior departmental chiefs, like Berlin's hedgehog, know one big thing[11] about decision-making – namely, that most of the discussion prior to coming down one way or another is much less important than, and hardly related in content, to the ways in which that decision will have to be justified publicly later on (March, 1999b [1996], 27): sense-making as powerful political narrative is much more important than anything else. As a result, they 'know' that investing in the quality of pre-decision analysis, understanding and discussion may do nothing for, and may actually undermine commitment to decisions made, and that, for the minister or the chief executive is far more important in getting anything done than that what is attempted should be entirely technically sound. Therefore, they would – if they read such things, which they do not, but their situation would lead them to believe this to be a general truth in any case – attend with particular care to the evidence that some advanced e-governance tools can actually erode commitment to decisions made following the recommendations arising from models and other such software tools (e.g. Landsbergen *et al.*, 1997).

The doubts and fears of the movement activists about the e-governance tools deployed by the professional policy analysts are the mirror image of those which the ambitious politician and the wily chief officer has. Where the competitive individualists climbing the greasy poles of politics and civil service or local government bureaucracy worry about too much rigidity and rationalism, the movement worries about too much fluidity. The fears of the movement are that the importance of principled action will be forgotten in the desire to understand very specific features of different policy problems, that the professional desire to fine-tune policy to peculiarly different gradients and balances of forces in each case or problem will lead to a loss of commitment to the overarching general principles that should inform policy, and which should not be compromised in order to design policies around today's empirically observed specific conditions. The movement shares with the leadership the importance of sense-making, but, seeing sense-making as a matter of subsuming problems and conditions under general conditions, it objects in a slightly different way to the idea that policy making consists in the making of a series of discrete decisions about what are seen as discrete problems. For the movement, if moral commitment applies anywhere, it applies anywhere. For them, to adapt Harold Wilson's famous description of the Labour Party, politics is a moral activity – in the extreme case, a crusade – or else it is a pointless and squalid process of fudge and accommodation. While the anxiety about loss of commitment to decisions from the use of the analytic e-governance tools to which the professional analyst cleaves is of real concern to the movement, the imperative is not felt here just to get something done at all: rather, the concern is that e-governance tools will fail to direct policy makers in doing the right thing.

Finally, the professional policy analyst too has her fears about e-governance tools of various kinds. Indeed, she even has doubts and concerns about the very tools of modelling, structured analysis scenario building, robustness testing and rational calculation to which, under the institutional pressure of her situation, she gravitates. The main concerns are the classical professional ones of efficiency and efficacy. For modelling system, the concern is whether the investment of money, time and effort in buying, installing, setting up and input of data into sophisticated software is worth the yield in analytical knowledge that can be achieved. For argumentation support and document storage and search systems, there are concerns about whether they will in fact reduce or increase the problem of intractability of data. Finally, there is a concern about whether the security and integrity of the proprietary information base will be compromised by the more open systems of information sharing required by some tools. However, she has greater concerns about the preferences for the use of and commitments behind e-governance tools espoused by those players in the policy process in contrasting situations. From the loose world of e-mail and word processing that is the practice of the isolates and of ministers and chief officers, she sees only chaos, indiscipline, lack of rationality, loss of knowledge, both insufficient exploitation of assets and actual undermining of policy capability through excessive individualisation. For example, Holtham (2000) has criticised the excessive use of e-mail and word processing and stand-alone storage systems by policy makers and managers in British government, and what he sees as the lack of commitment to the use of groupware and shared data systems. In the ministerial practice of information opportunism and the use of information systems to override detailed policy work, the professional policy analyst worries about the dominance of what she regards as 'symbolic politics' over sound policy design (Feldman, 1989). In the leaderless debates of principle by which movements work, she sees a lack of rigour and a lamentable lack of appreciation of the highly variable and specific conditions in which policy problems are presented, and an unrealistic failure to understand the varying powers and limitations of the actual tools of policy action available to the bureaucracy in implementation.

Both the selection and the style of use of e-governance tools, then, readily become the battleground over which conflicts are fought, that are really about rival institutional styles of conducting policy making, each of which is rooted in quite specific circumstances and situations, that are found in distinct niches among politicians and among civil servants and officers, and among the plethora of external players from individual experts in a particular policy or administrative area, through wheeler-dealer political consultants, by way of professional economic and environmental policy analysis institutes to campaign groups. Figure 9.2 summarises the argument so far about how each institutional style has concerns about the ways in which e-governance tools might be used.

Constraint ⇑	**Loose clusters of isolates**	**Central units, hierarchies**
	Futility	Excessive time commitment for gains achieved
	Rational decision procedures produce even more randomness in the real problems, through Goodhart's Law, the law of unintended consequences, etc.	Problems of unmanageability of data implications
		Loss of political discipline in decision making group due to openness of weakly constrained procedures

⇒ **Bonds**

	Competitive individualists	**Enclaves, sects**
	Excessive rationalism, rigidity in many tools	Excessive focus on the discrete decision, inadequate focus on whole principled strategy
	Dangers of loss of control of proprietary information to others who cannot be trusted with it	Loss of political control to technocrats
	Excessive focus on procedure, insufficient attention to opportunity	Loss of affective commitment to principled political action
	Crowding out of craft of judgment	
	Loss of political control to technocrats	
	Closure of political networks Loss of attention to public justification after decision	
	Loss of commitment to implementation through erosion of subtle, informal face-to-face signals	

Figure 9.2 What each of the institutional styles of organisation among policy makers worry about in the use of those digital support tools for governance that they tend to reject

Conversely, each style has its hopes, however limited (at least in the case of the isolated backbencher) for their peculiar preference in software support for their own work, and, if others were to adopt it too, the policy process in Figure 9.3 more generally summarises these.

If this argument is accepted, political judgment should be understood as both shaped and conducted by these conflicts of fears, doubts and aspirations about the ways in which other players in the policy process will seek to use e-governance tools to further their particular institutional style of judgment and sense-making. The hopes, fears and patterns of willingness to different systems flow from the styles of political judgment cultivable in each situation; in turn, those styles of judgment

Constraint ⇑	Loose clusters of isolates	Central units, hierarchies
	Occasional in-time alerting of erupting crises	Rigorous understanding of problems and instrumentation available to tackle them
		Oversight of, and therefore means of control over, dispersion of views within decision making groups
		Control over quality of data used in decision processes

⇒ Bon(

	Competitive individualists	Enclaves, sects
	Rapid communication	New ways to hold members to account to prior principles, and link their knowledge to those principles
	Identifying opportunities	
	Lateral thinking, opening new kinds of questions	

Figure 9.3 What each of the institutional styles of organisation among policy makers hope for, at best, in the use of digital support tools for governance

reflect the actual organisation of policy makers under each set of institutions.

Political judgment: disciplines and decision making as a ritual process

The chapter has argued that we need an organisational account of how information and communication work in policy making, and therefore of how the tools of e-governance are used. In doing so, it has challenged information technological accounts of how political organisations work. That is to say, there are reasons for rejecting the fashionable claims of those who suggest that in, and of themselves, the invention and adoption of information systems and technologies will wholly transform organisational practice.

On the contrary, people will use information and communication systems and banks of knowledge for purposes – whether conservative or transformative – that they already have. But those purposes and interests are not just given, somehow implanted at conception into the brain nor in early infancy nor from one's parents' occupational class, whatever the importance of such forces in the very long run. Nor do they flow in any simple or mechanical way even from being hired by a particular department nor from becoming a member of a party. Interests are produced, and with some difficulty, from many raw materials by institutional processes in the course of policy making. That is to say, interests are generated by the institutional styles, arising from people's situation in the policy process. Technologies are seen as successful to the extent that they are resources for people working in those institutional styles to further the ends their situations make available to them, or to the extent that they enable people to make settlements between the institutional styles that shape the work of the others with whom they have to deal in the policy process.

Second, it has been argued that the way in which the institutional styles work – and indeed, use technologies of e-governance – to shape political judgment is not, in the first instance, by skewing the content of discrete decisions. Political organisations working in the policy process are not primarily machines for making decisions: that is a feature that flows from their primary activity of making sense of their own experience and – with greater difficulty because of the pressures of their own situation as an organisation within the policy process – of the experience of those whose concerns they seek to articulate. Sense-making is the basis of political judgment. Allocating attention, selecting lines of causality to which to give emphasis, allocating responsibility and blame, framing problems, fitting choice opportunities and particular frames into the larger narratives that bind policy-making organisations together, taking and sloughing off responsibility for trade-offs are all activities that contribute to making sense of shared experience, and different institutional styles do this with different commitments that flow from the situations to which they are responses, and the sense they make of people's experience conflicts with the sense others will make: sense-making in the policy process is essentially and necessarily a conflict-ridden matter. Particular decisions are undertaken to make a contribution to this larger project of conflicting sense-making.

For example, long before CitiStat at Baltimore is about making decisions, it is about sense-making through defining individual accountability and telling stories about the city government in terms of individual's performance, which may be blameworthy or praiseworthy. The decisions that flow from CitiStat are the product of that sense-making, not the other way around. Similarly, in the DoE, the work of the modellers, the synthesisers of the outputs of the modellers and the policy brokers who make selective use of these outputs in waging 'battles of the models' are part of the weaving of shared stories about the world. The world comes to be seen as structured by the complex but essentially orderly interaction of forces which are amenable to intellectual analysis, and to action based on the gentle forcing together of sense-making around trajectories of development which have attracted ideally both authority and consensus, or, failing that, either one of them. Energy policy decisions flow from this capacity to represent the world, or to engage in what Vickers (1995 [1965]) called 'appreciation'.

Sense-making is not in the main best understood as a highly intellectual process of playing with ideas and arguments, in the manner of formal philosophy, for all that some would-be professionalisers of policy making in the high constraint, strong bond quadrant of the matrix, might prefer it to be. On the contrary, making sense is a rather mundane, quotidian affair, and one which is conducted through certain kinds of *practice*, which is often only articulated as formal argument later on. The problem with most accounts of policy-making organisation as machines for making a sequence of decisions – and therefore, of the account of policy making as in need principally of decision support systems – is that the practices in which sense-making gets done, rather disappear from view, or else they are misrepresented in narrowly instrumental ways as merely tools for making discrete decisions (Majone, 1989, ch. 1).

For example, many software and hardware e-governance systems are now available under the rubric of 'meeting management'. Most of them are designed to put great effort into those elements of judgment which consist in the generation of choice options and in coming to closure through some procedure that weights or aggregates the preferences of the meeting participants between the options generated. Fairfax uses a fairly straightforward tool of this kind in its Group Decision Support Centre. Many meeting management software tools are quite sophisticated in organisational terms, at least to the extent that they are offered as parts of packages that specify how meeting time is best used, how the physical meeting space should be arranged, and so on. Such systems have their uses, of course, but they are rather specialised. The account of political judgment offered here shows why. They assume that allocation of attention, the selection of lines of causation to emphasise, the allocating of responsibility, and the shouldering of responsibility are either matters that can be supported by entirely general systems that support forms of conversation, or else that they are conducted elsewhere. The result is that they provide some physical management of time and space and some procedure for a part of the sense-making process. But because this does not support the whole judgment process, the use of such systems can be curiously abstract, artificial and strained.

Most studies on how meetings work in organisations suggest three things are missed in the design of many of these systems, each of which is crucial for the argument of this book.

The first is that meetings are not principally for making decisions, but for making sense. More exactly, decision opportunities are occasions for sense-making in organisations, and for the institutional style or settlement between styles characteristic of the organisation to shape how people conduct the work of sense-making. Particular decisions are made in meetings – or at least as often, not made, postponed, ducked, or sense is made of shared experience in such a way that a decision is deemed not necessary – in order to further sense-making (Schwartzman, 1989; see also Forester, 1993, 1999).

The second is that sense-making is principally done in meetings, rather than by solitary cogitation, precisely because meetings provide a *ritual*[12] – or at least, ritual-like[13] – order through which sense-making can be done collectively (Schwartzman, 1989; Yanow, 1996). For institutions are not just abstract rules and norms and accountabilities: they are necessarily embodied in *material* facts about the layout of rooms and the gait, demeanour, and bodily performances of people for only thereby can accountabilities be communicated, sanctioned and fulfilled (Goffman, 1967; Mauss 1979 [1935]). It is unfortunate, and suggests quite the reverse of the truth, that the word 'ritual' is so often used in discussions of public policy to mean something empty or sham, engaged in cynically in order to deceive and mystify, or for the sake of good manners but without substantive content, as though showing something to be achieved through ritual were to unmask it as either pointless or else a kind of plot, and that the participants are merely dupes (e.g. Edelman, 1985 [1964], 1988; for a theoretical critique, see Douglas, 1966b, 1970; for evidence that participants in rituals of coming to judgment in public sector organisations are far from dupes, and that they reflect carefully on their roles and on the ritual they enact in meetings, see Westrup, 1996, 29ff).

On the contrary, meeting ritual in policy making organisations is what sustains the character and the structure of the relationships between people who attend them, and therefore supports the institutional style of the organisation.[14] Indeed, meeting ritual is but the focal point for many other rituals that govern relationships within and between organisations in the policy process even when those individuals are working in settings that are physically far apart. The codes that define how memoranda are prepared, how matters are prepared for discussion, how documents are prepared, how telephone conversations are conducted, when a meeting table is appropriate and when it should be dispensed with in favour of serried ranks or horseshoe or circular arrangements of chairs in different kinds of meetings, when whiteboards and flipcharts are used and where they are situated and when they are to be absent, also express the ritual form of the organisation and sustain the process of judgment, but it is generally in meetings that policy makers are able to perform rituals that give overarching focus and sense to the rest of their work (Schwartzman, 1989). The everyday rituals of contact between Members of Parliament working from their desks, the conventions of minute writing in the civil service, the organisation of circulation lists for copying, the etiquette for e-mail reminders of what documents have been placed for elected members in a database accessible on the local authority's intranet, are all subsidiary rituals to the great meeting rituals that are summarised in statements of Parliamentary procedure, the Osmotherly rules, the council standing orders, and so on. The most useful e-governance tools are therefore, those which are most carefully embedded in those rituals which perform these functions.

Sense-making, as conducted ritually through systems of meetings, is highly narrative in structure. Policies themselves, and the activities that comprise the making of policies, are conceived as stories, with different narrative structures in each institutional style. This is true at several quite distinct levels. First, in the framing of issues as policy problems, the most important early activity is the telling of stories and anecdotes (Forester, 1993; Weick, 1995, 2001; Yanow, 1996). Such stories are not merely casual instruments of persuasion, in principle eliminable in favour of rigorous quantitative analysis – as the favoured myth of the central professional unit would have it – but they are constitutive of the framing of policy problems and they are essential rituals by which attention gets allocated to policy problems: the focusing of attention itself is not merely done by telling stories, but consists in story telling. Second, the cycle of meetings, authorisations and approvals by which policy judgments are made as topics wend their way through bureaucracies, movements and parties and legislatures, or the pathways that are groped toward in what appear at the time to be meandering ways, are retrospectively organised into coherent stories, for which the sequence of rituals of judgment provides punctuation, structural clarity, and meaning.

Each of these features can be seen in the case studies. In Baltimore's CitiStat room, the now fully institutionalised ritual of the cross-examination and the careful layout of the room to sustain the highly adversarial order of the meeting are at the heart of the whole programme and its impact upon the culture of the city government. When Mayor O'Malley insists that GIS images shown in the CitiStat room show the names of area managers, he is expressing the individual accountability institutions that his administration stresses in palpably material ways.

The meaning and the efficacy of e-governance technologies such as GIS and the other off-the-peg applications used in CitiStat are both fully embedded in this ritual order. Similarly, the ritual order of the horseshoe and the facilitator supported but otherwise generally internally egalitarian organisation of deliberation in the group decision support centre is precisely what sustains the usefulness of the meeting management technology. As an exercise in political judgment, redistricting debates in Fairfax's GDSC room were made possible by the ritual order imposed by the discipline of facilitated computer-supported meeting management. The ritual order of interagency meetings in federal government is more conventional than either of these. What is distinctive is the ritual character of the interactions running over many, many meetings by which models are used to 'refine the debate'. That refinement is achieved, not as a mechanical consequence of cranking out results from models, but from the iterative process of interpretation and argument about model results to isolate which assumptions are at once sensitive, critical and negotiable and amenable to some kind of political settlement between the agencies. Each of these rituals supports the possibility of sense-making through narratives. These are respectively, tales of villainy and heroism that make no sense outside the ritual order which defines performance management in the adversarial terms of CitiStat; stories of collegial commitment to proper representation of all the geographically diverse areas of the scattered communities of Fairfax, which are given concreteness in the use of GIS; and narratives of both politically and intellectually responsible and credible cogitation to produce government policies that can find some settlement between the rival tribes which populate the energy policy world, which only have meaning to the extent that the ritual conciliation process between rival models can be built as a microcosm of wider political and intellectual conciliation.

In the first instance, then, to understand how e-governance tools either are or at any rate can be embedded in such ritual or ritual-like forms, we must look to the ways in which the different institutional styles select and use rituals, to sustain their peculiar sets of organisational relationships between policy makers. What kinds of rituals can each most readily use to sustain its peculiar pattern of relationships and style of judgment?

Consider first the central professional units. Here, time in the policy process is at its most structured, for it is organised by the cycles defined in both the disciplines of analysis and by the cycle – however mythical this is as a *comprehensive* statement of how policy works as a whole – of formulation, legislation, implementation, scrutiny, reformulation, and by the discipline of political time over both the parliamentary annual cycle and over the electoral cycle. Judgment here consists in analytical inference on the basis of high levels of trust permitting information sharing between defined groups of peers, with much less trust toward outsiders. Therefore, e-governance tools here are most useful when they can readily be embedded in a highly regular cycle of meetings, each of which is governed by defined procedures of judgment – both sense-making and decision making – which encourage and support the sharing of information and the mutual disciplining of analysts in the quality of their performance of their craft. This explains why, for example, the highly ritualised forms of chauffeured decision conferencing used by the British ICT company to work with local authorities in budget setting, are often most successful in these settings. Again, internal committee systems that oversee analytical and policy development work well in this setting, and lead to the creation and use of highly structured documents bringing together models, evidence and recommendations on the basis of which formal consultations are undertaken to elicit and enable the analysis of comments on the proposals. This last shows why the narrative and ritual style of this setting tends to be rather juridical. In juridical contexts, problems are presented and attention is directed from outside, by litigants. In the same way, for the central professional unit, problems and attention are determined by the priorities of political and administrative leaders, by the harvest of the intellectual disciplines of policy work they use, and by pressure from movements and other semi-isolated individuals who are in respectively constant and casual communication with the policy staff. Just as in juridical judgment, evidence is gathered either inquisitorially or adversarially – both highly ritual forms of judgment – or in both manners in sequence, and there is a structured narrative by which closure is reached, responsibility allocated and shouldered. E-governance tools are most welcome here, that are embedded in rituals of judgment which permit such clear allocation and closure, and chauffeured decision conferencing is a good example of this, but some other meeting management tools also make sense here, precisely because what those tools take for granted sometimes actually can be taken for granted in this setting, although this is not true for the policy process as a whole. It is for this reason that in such hierarchical contexts, political judgment is *process oriented*, in order to provide structured understanding of problems.

For ministers, permanent secretaries and chief officers, the rituals of judgment that make most sense are rather different. Their experience of time in the policy process is disjointed, structured by their efforts to seize temporary advantage and proprietary control over information at the expense of their rivals, and judgment is a matter of cultivating and exercising insight in order to identify opportunities and comparative advantage over other ministers or departments vis-à-vis the Treasury or over other parties, or factions. For them, rituals of judgment must enable them quickly to raid banks of information to pick out the valuable pieces, and ensure their control over these, and to see the 'big picture' that enables them to use sense-making for larger political advantage. Therefore, they require rituals that advantage *fast learners* in the judgment process, that support them in competition for authoritative control over information and knowledge, that move very quickly to closure, and that enable them to focus on the allocation and shouldering of responsibility (cf. March and Levinthal, 1999 [1993]). Their patience is therefore limited with the rococo carefulness of elaborate modelling and consultation. For them, it is important that there are quite distinct ritual opportunities at the beginning of such processes in order that they can allocate the attention of those they lead, and toward the end of those processes, they can raid the results. That these will be felt as a kind of violence to the integrity of the careful analytical judgment rituals of the professionals is simply part of the inevitable conflict of institutional styles which organises the policy process. Moreover, because too much sharing of information and too much revealing of one's own preferences is dangerous for them in the game of information poker they must play, they are often less than enthusiastic about those meeting management and decision conferencing tools that too explicitly display and represent the balance of opinions of rival competitive leaders in the same room. A certain studied vagueness is essential to the leadership role, and they prefer rituals of judgment that enable them to maintain it. E-governance tools therefore play their most important role for this group in the periods when they are preparing for these special meetings for priority setting and reporting back. Thus, their situation explains their preference for individual e-mail, for systems of etiquette in electronic conferences and argumentation support systems which permit them to be silent 'lurkers' for a while, but then to make authoritative statements of the allocation of attention and responsibility and statements of closure, and for rituals of judgment in which they can secure prior and privileged access to the products of the professional judgment process, are explained by their situation. Their narrative style is heroic, rather than judicial: the policy process is understood as stories of individual political entrepreneurs heroically overcoming their rivals to break through administrative gridlock and vicious cycles of causation in policy problems with astutely insightful, highly visible and specific measures of government action. Therefore, political judgment here is *backward-looking* in the context of devolving blame but *forward-looking* in the 'heroic' context in which the individual leader seeks to use judgment to impose closure on the sense-making process to define something that can be recognised as a decision.

By contrast, the rituals of judgment in face-to-face and in online communication within factions and movements are quite different. Time here is indefinitely structured: lacking control over commitment to action, there is not the same urgency that structures time for the leadership nor the cyclical disciplined time of the professional unit, and judgment is concerned with the appropriate application of principle. Therefore, rituals of judgment can afford to be more generous with time, and more deliberative in character. However, the precarious position of leaders means that it is hard for them to use authority in rituals to allocate the attention of others, to force closure, or to use the collective judgment process as a resource that they can raid. They must submit to the combination of elaborate but fragile politesse of protracted egalitarian deliberation which can so easily flip over into rituals of blame and accusation for alleged failure to conform to the principles to which the movement is dedicated. However, coming to closure around principled action is of vital importance, but that closure must take the form of a consensus of the true believers who are prepared to stand up and be counted. The narrative style of rituals of judgment here tends to be salvationist: the process of coming to judgments should reflect the saving power of moral commitment and their triumph over the forces of vested interest, bureaucratic compromise and muddle, while at the same time exposing the false friends of principle as trimmers, slackers or (in extreme cases) traitors. The etiquette of the Usenet world, with its flaming and mutual admiration, then, is but one part of the rituals of judgment required: the process of coming to closure is best achieved in some form of face-to-face meeting, which may be supported by hardware and software, provided that the ritual in which the software is used, is situated in a narrative that allows for the triumph of principled consensus. Here, therefore, political judgment is in the first instance *backward-looking* in order to mark off the past and present period for which the appropriate stance is critique and blame from the more vaguely defined future.

Finally, the rather shapeless experience of time in the policy process gives the isolate backbenchers and their administrative analogues a picaresque form to the narrative of their relationship to judgment, and no

Constraint ⇑	Loose clusters of isolates	Central units, hierarchies
	No distinctive ritual style	Facilitated, structured procedures, e.g. decision conferencing
	Time: shapeless	
	Narrative style: picaresque	Structured cycles of meetings and hierarchies of committees and special commissions
	Immediately oriented judgment is key: find a way to get through the day's urgent crises	
		Time: cyclical, regular
		Narrative style: juridical
		Process-oriented judgment is key: develop an adequate and rigorous understanding of the structure of the problem
	Competitive individualists	**Enclaves, sects** ⇒ **Bonds**
	Loose exchange on episodic basis as initiated by particular political entrepreneurs, until blocked by others	Collective consensus-building with minimal leadership; in near-permanent session, until great defining moment of consensus is symbolically achieved
	Time: episodic	
	Narrative style: heroic	*Time*: indefinite, but ruptured
	Immediate, forward and backward looking judgment are key roles: allocating attention, allocating blame, both devolving and shouldering responsibility	*Narrative style*: salvationist
		Backward-oriented judgment is first priority: identify responsibility for problems that arise

Figure 9.4 How the different institutional styles of organisation create rituals to exercise judgment

distinct or typical ritual form of their own. Rather, isolates tend to be the essential audience, foil, arena, ballast, and half-heartedly supporting cast for the rituals of judgment of the three more active institutional styles of policy making.

Figure 9.4 summarises the argument of this section.

We have already seen how judgment is intimately related to learning, and differently so by institutional style, and it was noted above that among the competitive individualists of the leadership, being a fast learner, or being able to make *rapid* sense of the available pool of experience, is among the most prized and cultivated skills in political judgment. Recent formal modelling work on organisational intelligence has shown that it matters very much how the prevailing institutions within organisations structure relationships between slow and fast learners: both are valuable, and excess of either in an organisation can, it seems, bring dangers (March, 1999c [1991], March with Lounamaa,

Constraint ⇑	Loose clusters of isolates	Central units, hierarchies
	Speed of learning is largely irrelevant. They both get surprised when what they learn turns out to be wrong and when it turns out to be right. But on balance, fast learners are slightly more likely to spot ways to cope with the random hazards of political experience. Rituals of judgment are of little help in enabling this	There needs to be a balance between fast learners and slow learners. Fast learners tend to pick up on new ideas, which – by the law of averages – tend more often than not to be wrong. However, slow learners are often – but not always – right. Cleaving to old but relevant wisdom is sensible. The onus of proof lies with the fast learning innovators. The role of rituals of judgment is to ensure this balance.
	Competitive individualists	Enclaves, sects
	Fast learners are the most valuable in a decision making group, but also the most dangerous competitors for other fast learners. The role of rituals of judgment is to enable the fast learners to emerge and to use competitive pressure to get the best from them.	Fast learners should be distrusted: they tend to lack loyalty, and have only tenuous commitment to the enduring principles. The role of rituals of judgment is to discipline the fast learners to the enduring principles that slow learners rightly cleave to.

⇒ Bon

Figure 9.5   How the different institutional styles of organisation treat the relationship between slow and fast learners in rituals of learning

1999d [1987]). This is exactly what we should expect. For rituals of judgment function, among other things, to give emphasis to the speed of learning that particular institutional styles, each with their peculiar strengths and weaknesses, are committed to, and the experience of political and policy time that is their particular situation. Without labouring the point, Figure 9.5 summarises the differences.

E-governance tools will then be selected by each of these groups that privilege the learning and the learners at the pace or balance of paces they favour.

Settlements and mixed cases

The four institutional styles that have been described here might be thought to be simply stereotypes, or simply four situations picked at random. However, there is good reason to think that they are neither. That the description has stylised them slightly is certainly true.

However, I believe that they are not simply stereotypes, if by that one means that they systematically falsify the situations they seek to describe. Indeed, any reader of the more candid political memoirs and biographies of ministers, or any reader of the more candid descriptions of life in the Whitehall 'jungle', as the cliché of broadsheet political journalism has it, will recognise the characterisation I have offered of the position of the leadership contender, the departmental permanent secretary or the agency chief. Likewise, readers of the writings of movement activists, whether of left or right or of consumer or green suasions that would resist positioning on that simple axis, will, I believe, readily recognise the situation and the outlook of people in that position. Those who have studied the writings of the post-war technocratic policy advisors who sought to put policy analysis in government on a more rational and scientific basis (e.g. Simon, 1997 [1945], 1977 [1960]), and sought to draw upon each succeeding generation of computing power to support their aspiration, will surely recognise my description of both the situation and the commitments of the central professional units.

More to the point, these four are certainly not simply picked at random. They are exemplary of the situations of other players in the policy process, for the information, influence and decision problems and aspirations of *all* those who seek to work in or upon the policy process are profoundly shaped by the constraints of rule, role and given fact at one end or the more relaxed constraints of the voluntary choices of others at the other end of one spectrum, and by the degree to which bonds to one's peers are tightly bound by group *esprit de corps* or shared purpose, or else loose because of isolation or competition on the other spectrum. To casual observation, the situations of the commercial lobbyist, the advertising agency selling its services to the party for the design of election campaign posters and broadcasts, the hack speech writer, the senior civil servant, the front line police officer or school inspector, the agency chief, and so on may all seem so entirely different from one another, that it might seem pointless or misleading to offer an account that explains something of how each will want to use information and communication systems in their attempt to influence governance. However, if we are to understand how software tools and network hardware are now and might in future be used, we cannot rest content with raw description: we must try to find explanatory forces. These two dimensions represent a summary of the big shaping forces upon which social science has converged over a century.

Moreover, it becomes clear from the way in which these two dimensions are cross-tabulated, that, at least at this more general level of understanding, the four categories are, in a certain confined sense, exhaustive.[15] There can be degrees of constraint and degrees of tightness or looseness of bonds to one's peers and thus there can be degrees of realisation of each of these four institutional styles, but there is no Archimedean position from which one can shape the policy process that is not subject to these forces in some degree or other.

However, in another very important but also restricted sense, the four positions are not exhaustive. For there are many organisational situations in the whole mix that makes up the policy process in any country that represent combinations, mixtures, or, more relevantly, *settlements* or *coalitions* between two or even three of these institutional styles. The four institutional styles exhaust the possibilities for, as it were, organisational raw materials from which practices of governance and e-governance might be assembled, but there are, arithmetically, six possible coalitions between two styles, and four possible settlements between three of them (Thompson *et al.*, 1990). Finally, there can be and there are often settlements in which something is given recognition and institutionalisation from all four. Each of these eleven possibilities can vary in the extent and degree to which any and all of the four basic institutional styles are represented. In this sense, then, there is extensive scope for change and variation with these four basic institutional styles.

When we shift our lens back from the particular situation to the panorama of the policy process, we can see that in practice, departmental spending ministers do manage to rub along with their professional staffs, find some accommodation with the plethora of factions within their party and in the policy fields their ministry must cover, and do in practice manage to respond to the immediate concerns the back-benchers press upon them, and conversely for each of the players. Over time, differently in different administrations, power shifts between these players in a kaleidoscope of different settlements. Indeed, it has been argued that they are *dependent* on each other, despite their conflicts.

However, it is in the nature of settlements and coalitions that they are never free from tension and friction, and that they are often temporary. The fears and concerns about e-governance as conceived by the others may stand representative for the many ways in which the four basic institutional styles continue to *disorganise* each other, even as they find ways to rub along together. For their own internal organising principles are the very ones that most disorganise the others. The very source of cohesion is also the source of conflict, just as the very source of inclusion for the professional unit and the movement is also the source of its exclusion of outsiders. Sometimes this process of disorganisation comes to a head, and some settlement breaks down: bitter conflict erupts within a party between movement rebels and the leadership; or relations break down between the media-savvy showpersonship of a minister and the purely analytical and professionally oriented judgment of a government department; the isolate backbenchers gradually drift away and a party suffers a crisis of political recruitment; or the movement finds that as its activists 'burn out', they become isolate backbenchers and its problems of schism become even worse within the group remaining, as they now lack the ballast of the periphery of semi-isolates that held them together with the always slightly illusory sense of mass momentum. In a certain sense, the policy process only works through settlements and coalitions. Yet for the same reasons, these settlements are inherently fragile.

It is open to anyone to say, of course, that one of the four basic situations is the correct one, and that everyone who is a player of any kind in the policy process ought to try to make their situation more like this one. There is, after all, no shortage of advocates for the professionalisation of policy making as a rational and technical process, just as there is no lack of those who would advocate that the principle of one kind or another ought to decide all. Advocacy for the important role of leaders cultivating and exercising insight and brokering decisions has never been lacking. Perhaps the only institutional style that has few advocates is that of the loose clusters of isolates, although their voices are not entirely unheard, although more often at second hand or in private than in the form of direct advocacy.

Nevertheless, as the tenor of the argument in describing these four may have suggested, such straightforward advocacy for a single institutional style is at one level misplaced, and at another level misguided. If we are considering the short run, then it is misplaced to imagine that it is at all easy for people to resist the pressures of the situation they are in. If we are considering the longer run, in which it might be possible to change that institutional setting and to increase or reduce the strength of authoritative constraints or of bonds to peers, then such advocacy is misguided, because each of the four has both advantages and disadvantages. The blinkers and the weaknesses of each are the converse of those of its neighbours in the scheme.

Again, someone might say that any but their least preferred one of these positions is unacceptable, and that we ought to do whatever we can to reduce or eliminate the occupation of a particular quadrant of the matrix I have used. But the same considerations show the weakness of three-way settlements.

For the importance of the way in which each institutional style disorganises the others shows that any attempt to impose upon the policy process as a whole, a paradigm of professionalisation, of leadership brokering, of principled accountability to vigilant movements, or even of day-to-day coping and survival, will soon prompt a contrary response by the other institutional styles, which will reassert themselves in reaction as people express and organise their frustration, born out of the situations they will sooner or later find themselves in, with such a dominant style of information and knowledge management and of decision making.

The same considerations about how each institutional style will re-emerge in reaction to the over-assertion of the others suggests that, for all the fragility of settlements and coalitions between institutional styles, it makes more sense to think about settlements that reflect and give recognition to at least something of concern to *all four* institutional styles. Whatever we might try to do by way of conscious and deliberate effort of organisational engineering to arrange the relationships in the policy process, sooner or later, all four institutional styles will re-emerge, and therefore, any viable arrangement must give some recognition to each (Thompson *et al.*, 1990). Judgment, on this understanding, is the work of the fox who knows many things, not the work of the four hedgehogs, each of whom knows but one big thing. The design of such four-way institutional settlements, albeit that they will be fragile and that new settlements will constantly have to be assembled as previous ones collapse, is, on this understanding, the central challenge for the public policy process.[16]

Something of great importance follows from this, about political judgment and the kinds of tools that policy makers need in order to exercise it. If the central challenge for public policy is to try to assemble settlements, fragile and provisional as they will be, between these institutional styles in the policy process, then political judgment must be measured by the degree to which it contributes capabilities to this process. However, the argument clearly shows that there is nowhere outside institutions that politicians and policy administrators can stand in order to do this. They must work toward settlements between styles from wherever they are situated, and from the styles or combinations of styles that govern each of their particular settings. All five elements in political judgment are called into play by this challenge, for it is essentially a challenge to undertake collaborative sense-making in conditions of ineliminable conflict. Conflict resolution is then at the heart of political judgment. Therefore, when we are looking for information and communication systems that will support policy makers in their central tasks, we are looking for sense-making support systems (Weick and Meader, 1993), before we are looking for decision support systems, and we are looking for ones that at least enable and ideally, implicitly or explicitly encourage conciliation of a kind that enables all players to give some recognition in settlements to at least some of the hopes and fears of each institutional style.

10
Ways Forward

This chapter considers how policy makers might think about selecting the e-governance tools that are of the greatest use to them, given their particular situation, in order to help them to sustain political judgment, and also offers some considerations that the information technology industries supplying e-governance tools might want to bear in mind when designing for and selling to policy makers. The second half of the chapter is concerned with the wider issues of public policy that are raised by or in association with the growing use of e-governance tools, in particular for the forms of accountability in which their use is embedded. Finally, the chapter considers the question of the larger relationship between e-governance as something which can be used by policy makers to sustain their own political and policy judgment, and democratic renewal, which calls both for the political judgment of policy makers to be held to account and to be responsive to the popular will, but also requires that popular judgment should also be supported, and – subject, of course, to important safeguards – cultivated.

Perhaps it is helpful to begin by considering the implications of the argument of the book for the question, 'what might be reasonable expectations from the use of e-governance tools?'. The present argument has been that from the use of the technologies *alone*, we should have quite modest expectations.

Take what must be one of the simplest case of trying to use technologies as an aide-mémoire to reduce confusion in a meeting. Schwabe (2000) describes the Cuparla initiative for electronic meeting support for local authorities, as it was applied in Stuttgart City Council. The scheme provided elected members with a variety of technologically fairly straightforward kinds of aids and the services of facilitators with which to make better use of the systems offered. Noting the complexity of many votes on alternative amendments to resolutions, Schwabe reports that many councillors become confused during meetings when amendments are tabled orally, as the agenda papers provide them with only those tabled in advance: the result is that sometimes members themselves had to wait until they saw the minutes prepared by the staff to understand what they had decided. Stuttgart councillors are by no means alone in their plight: members of most national parliaments could report similar stories. 'If the individual councillors' computers are connected to large shared displays,' he writes (2000, 387), 'the fine-tuning of decisive texts can be finished during the meeting. Each member publicly displays his or her version or changes the publicly displayed text until the council converges to a shared decision' (see also Schwabe and Krcmar, 2000).

That these tools *can* be used to eliminate confusion, perhaps even simplify meetings, at least ensures that everyone present has information about the draft resolution being debated, in the way that Schwabe claims here, is not in doubt. But whether the technologies *will* lead to these happy results is a function of the degree to which the city councillors of Stuttgart, in this case, are prepared to submit to the disciplines and rituals of working in new ways, to working together much more, and to use the changed ritual order of using a huge screen as the central focus of their chamber in order to manage their individual and collective attention in this way. If, however, it proves to be in the interest of some councillors – no doubt, not in upright and noble Stuttgart but in many less decent places – to get their way by exploiting precisely the confusion that can be generated by tabling oral amendments at the last minute, and indeed by tabling several of them in such a way that they can obfuscate debate, filibuster, distract their fellow councillors' attention, and so dominate the proceedings, then they are likely to find ways either to dispense with the screen when it suits them, or else to find new ways of subverting the Schwabe's aspirations for the use of these systems, even while keeping the tools in the chamber. It will not be beyond the wit of the determined and Machiavellian councillor, if Stuttgart should have any such in future, to find a way so to clutter screens with amendments as to secure confusion if it might suit their interests. Schwabe and Krcmar (2000) also found that greatest satisfaction among decision makers with their group decision support technology and discipline is reported when the physical setting is unfamiliar: this suggests the possibility that as councillors grow accustomed to the system and the discipline of meeting management, their attitudes may mature and change as they learn how to 'work the system'.

If this dependence of technology upon disciplines and the nature and degree of collective working afforded by the prevailing institutions, and the interests that those institutions define for particular people (e.g. individualist poker players) can be observed even for relatively modest goals for the improvement of what might be considered 'quality' in decision making, then it is even more clearly the case for goals of greater ambition.

It does not follow, of course, that e-governance technologies are not important, or not effective. Rather, they can be important and effective only in the context of institutional processes – and, where these are missing, institutional changes – that sustain ways of enabling poker players, chess players, football players and snap! players, to collaborate to some degree, even though they are never going to play the same game.

When, for example, the US Department of Energy officials speak of modelling 'refining the debate', this represents for them a genuine and important gain, if one that might be thought modest by the standards of true chess playing technocrats, and one that justifies for them the expenditure, the development of capacity and expertise that is entailed. The ritual order of the interagency meeting, the institutionalised expectations about how 'battles of the models' are conducted, the roles of the different energy policy layers, and the interests that these roles and institutions define together create a context in which the technology can, for example,

- structure the focus of attention;
- act as an aide-mémoire for making certain claims or putative facts more salient than others, including putative facts that make up the collective memory of the policy making organisation;
- enable people to identify the most critical and controversial assumptions, on which more work either of an advocacy or research nature, is required ('refining the debate');
- probing the degree of commitment to a candidate decision;
- provide a forum for debate that in turn calls for the recognition of a certain ritual order in the space defined by the presence of the technology; and
- provide a means of speeding up the communication and aggregation of preferences, and the development of what-if accounts based on extrapolation from actions being considered (cf. Fairfax County's use of its group decision support centre, discussed in Chapter 6).

None of these things is always and everywhere an unambiguous gain in 'rationality', if by rationality we mean the technocratic rationality of the researcher or even the civil servant. Greater speed, for example, can support many kinds of goals in a debate including the most political, just as slowness can support either greater deliberation or simple delay and filibustering. However, in many contexts, these will be important achievements and worthwhile.

Introducing the technology first as the focus can sometimes be important as a rhetorical act, in order to help persuade people to adopt the ritual that makes the technology most meaningful. Mayor O'Malley in Baltimore, for example, partly promoted his CitiStat system on the basis that it was making the best use of technology to manage data for performance, even though the really important driver of the system is the ritual order by which accountability is impressed upon agency managers. However, it would be wrong to assume that this is the typical case. In Fairfax, for example, the commitment to the wide use of GIS technology, rhetorically important as it is, seems to have followed as much as preceded the commitment to think spatially about resource allocation or to develop more collegial styles of working, although the availability of the resource has also reinforced this policy-making 'thought style'. Introducing technology is by no means a failsafe way of kickstarting organisational change, as the many studies on failed IT projects attest. Most reviews of the literature on organisational change regard technology as something secondary, something to be prepared for by work on organisational processes (e.g. Fountain, 2001; Proehl, 2001; Burke, 2002).

If the argument of this book is accepted that political judgment is something which is negotiated, neither something possessed nor something cultivated by simply working harder at using probabilities and analysing consequences, then it makes little sense to pursue the goal of cultivating political judgment by technology alone. If it is accepted that there are several styles of political judgment, but not infinitely many, and indeed only basically four which can be combined in various hybrids, then the sensible goals for the use of e-governance technologies to support it, are

- to support and strengthen the information-using and information-rejecting capabilities of *each* of the basic styles which will spring up in any political system; and
- to support disciplines, rituals and uses of technologies that can support better *negotiation* between them so that they can collaborate at least to some degree, rather than simply produce gridlock or guerrilla conflict that will weaken the policy process, whether the technology

in use is that of pen and paper or the most advanced artificial intelligence systems.

This chapter now considers each of these goals in turn.

E-governance technologies and institutional styles

We must therefore ask next how the analysis offered in Chapters 8 and 9
helps us to think about how policy makers in each situation and subject
to each of the four institutional styles might further the objectives that
their situations lead them most readily to adopt, by investing in and
using particular types of e-governance tools. This can be, of course, no
more than a first stage in thinking about how e-governance can and
should be developed, for, as the concluding sections of Chapters 8 and 9
argued, although it is not easy to break out of these institutional styles,
we may have reason not to privilege these situationally grounded
objectives. In this section, then, the challenge is to link up the range of
e-governance tools identified in Chapter 4 with the institutional styles
of political and policy judgment characterised in Chapters 8 and 9.

Central professional units

This task is easiest for the professional central units and their staffs in the
civil service, in policy divisions of local authority departments, in health
authority executives, and so on. For the information technology industries have already been more assiduous in designing systems that fit their
training in formal disciplines, their ability and willingness to trust their
peers with information, their collaborative structures, their articulated
divisions of labour, their style of learning. There are many group decision support systems, argumentation support systems, electronic meeting management tools, problem structuring tools, expert systems and
neural nets and modelling tools that fit well with the capabilities and
requirements that flow from their situation. It is unsurprising, therefore,
that it is among these groups that the adoption of the e-governance
tools currently available has been fastest and most widespread.

Almost certainly, there is scope for both wider and deeper use within
the professional policy analysis centres of the departments of state,
agencies and local government. Of particular importance in the next
few years may be the coordination and integration of the uses to which
different e-governance tools are put, within these central professional
units.

There are at least dimensions of coordination that are relevant here. The first is concerned with coordination and integration of the processes of policy making horizontally between departments and agencies (6, 1997a; 6 *et al.*, 1999, 2002; Strategic Policy Making Team, 1999; Bouckaert, Ormond and Peters, 2000; Performance and innovation Unit, 2000a;). In particular, there are good arguments, based on the general case for more holistic governance, for

- cross-agency and cross-departmental groupware, document profiling and document flow management tools that operate between agencies, in order to provide better structure and control for inter-agency consultation and joint policy development;
- greater use of argumentation support, organisational memory capture, deliberation support, and meeting management tools to facilitate cross-departmental policy making activity;
- greater investment in the development of formal models of policy problems, in which the variables used range over causal forces that are the focus of policy attention across government; many formal models are developed within single departments and reflect mainly the focal concerns of those departments.

However, greater coordination and integration is required between the policy-making activities for which the different tools of e-governance are used within and between departments and agencies. The point is certainly not that there could, still less should, be a standard series of stages in which these tools are used: in any policy-making process, the sequence of basic activities involves repetition and returning, leaping over activities that might seem to an outsider 'logically' to be expected to be undertaken at a particular point, and, contrary to the aspirations of rationalists, policy making is usually the more robust, flexible and responsive for this.

However, thought needs to be given within the executive to the ways in which, should certain common eventualities occur, procedures involving the use of particular tools of governance can readily be put into place – for example, the sudden defection of a department or a minister from a previous consensus, a shock given enormous media attention that makes a policy under consideration impossible to legitimate, or the sudden coming to light of evidence about implementation and capabilities that undermines the current collective train of thought. In such situations, it is often more important to focus on the tools that support processes of articulating and conciliating between preferences, than the tools that represent so-called 'hard' causal variables defining policy problems, yet the trained instinct of many policy analysts working in central professional units is to reach for the latter.

Second, given the speed at which policy analysts often need to move between activities in the policy process, there is also a need for investment in groupware systems and intranet systems that are so designed as to facilitate the greatest ease in moving between tools of e-governance, including reaching out from within one to another, to support the greatest flexibility that is consistent with the intellectual disciplines and the political accountability as well as the peer group structures that define the central professional units.

Competitive leaders

The e-governance tools that ministers, permanent secretaries and agency chief executives can use and want to use are very different. For no group in the policy system is the finding of the last generation of e-governance studies truer, that attention is their scarcest resource. For their attention must be spread thinly over huge numbers of problems; therefore, what they need and can use is not more information, but less and more relevant, better selected information, for which less time is needed in searching and the significance of which is easier to appreciate (Simon, 1977 [1960], ch. 4). Fast learning, but without being seen to be in need of learning, is the priority for this group. Moreover, these are the people for whom time away from the computer screen, in face-to-face conversation, is often most important, if they are effectively to carry out their key role in sense-making (Weick, 2001, ch. 20, esp. 454–6).

This fact suggests that one key e-governance technology that should provide greatest assistance to this group is the intelligent digital agent. For agents are essentially neural nets that can be 'trained' inductively to search, select, analyse, précis and organise information to meet the preferences and interests of the user. The problem with digital agents is that one needs to take time initially to 'train' them, and many busy leaders in the policy business cannot easily make that time. However, of all the competing priorities, there is a reasonable argument that in many cases, this may be time better spent than in at least some of the meetings with which it is all too easy to fill up a minister's diary, or for that matter, a permanent secretary's or a chief executive's.

However, on the other side of the same argument, nothing is more dangerous for a competitive leader for her or him to forgo maintaining her or his access to information resources that may only infrequently or irregularly be required. Many traditional management information systems provided a series of headline indicators on problems or processes, but allowed 'drill down' to more detailed and less aggregated or less summary information on request (McNurlin and Sprague, 1998, ch. 13). These still have their place, but provide greatest value when their search and analysis capabilities are integrated with capabilities to search and utilise information from across several networks, and not only the proprietary ones, while keeping proprietary information secure.

Similarly, as the volume of e-mail rises to silt up the 'in boxes' of these leaders, they will need to make use of filter systems that sift through their messages automatically, select some for simple automatic acknowledgment, others for referral on to departmental staff or elsewhere, some for one-month turnaround by administrative and secretarial staff, and others for immediate and urgent action by the leader to whom they are addressed. The US presidential site at http://www.whitehouse.gov offers a web mail facility to US citizens who wish to send a message to the president; behind the site there is a highly sophisticated automated e-mail response system, which analyses and mines much of the personal data provided by correspondents as well as filtering the content of their messages. Experience in the US Congress suggests that as politicians, in particular (because of their direct democratic accountability), introduce such systems, they will be under pressure to make public, perhaps on their web sites, the rules by which their e-mail management software is instructed to sift and sort incoming communications, and to take responsibility for the consequences of the decisions their software makes.

Agents and filter systems, however, are essentially tools for the improvement of listening, which is but one element in the sense-making process for which competitive leaders are responsible. This group also needs to speak, to tell stories, to try out new stories, to strike settlements to conciliate in conflicts with their peers and their rivals (often the same people), to sound out those rival peers in the hope of getting them to reveal information while keeping their own most valuable or vulnerable information to themselves, to associate themselves publicly with credit for successes and to devolve blame quickly for the inevitable failures. The activities of sounding out rival peers, conciliating, and trying out stories are ones that sometimes lend themselves to the use of electronic meeting management, deliberation, organisational memory capture and argumentation support tools. Of course, where the risks are simply too great in the disciplines of openness that such groupware calls for, where one will be forced to reveal more than one wants to, then one can only expect that competitive leaders will prefer to resort to e-mail, private videoconferencing and individual word processing (Holtham, 2000). Nevertheless, in conciliation, these more collective tools will have their uses even for competitive leaders, for they can be used to create arenas in which deals can be made on a multilateral basis, where peers can hold one another to account, and where coalitions, however fragile over the medium term, can work effectively to sustain enough cohesion for the system to work. If the holistic governance agenda is, as is widely argued, increasing the pressure of public expectations upon these leadership groups to demonstrate that despite their rivalries, they can bring greater integration and coherence to public policy and service design, then these more collective e-governance tools could be used to support increasingly important processes of coordination in policy making at the most senior levels.

Party groups within the legislature and principle-based movement agencies within the governance system

It is of course beyond the scope of the present study to consider how far it is for government to enable factions and movements outside the public sector system of governance; instead, the focus here is confined to those policy makers within the legislature and the executive whose situation resembles that of the classical faction or movement, in their organisation around specific commitments of principle, their weak leadership but strong bonds between members. These include certain regulatory bodies committed to the enforcement of laws on civil liberties, non discrimination and similar rules based upon moral principles.

A key challenge for these groups and bodies is to obtain otherwise proprietary information from those they regulate, challenge or seek to hold to account. The question of how far information that they request should be available to them, online or otherwise, and at what point in time in the policy process, is never an easy question to resolve, and it has been extensively debated in the context of recent debates about freedom of information legislation in Britain and elsewhere.

However, like the central professional units, organisations in this situation need to sustain their consensus and both to deepen and spread among their members their appreciation of the implications of the principles to which they are committed, in the changing contexts of public policy and circumstances. For this purpose, argumentation support tools, electronic whiteboards, problem structuring tools, and meeting management systems can be used to play important roles.

Isolates

For backbenchers outside factions, policy staff and executives in the more marginal agencies, and some other similarly situated people in the policy process, the key resource for which any e-governance tool will be appraised, is its capacity to assist them in finding coping strategies, and in making best use of the rare and unpredictable cases when access to influence suddenly opens up to them for a brief period. In keeping with the general eclecticism with which people in this situation must operate, there is no reason to expect that any particular type of e-governance tool will have any greater appeal to them than any other, although the weak peer bonds between them make it more difficult for them to see much point in using, on more than an irregular and opportunistic basis, of the tools that require more sustained collective collaboration and trust.

Mixed and unstable settings

Select Committees and scrutiny committees can, as we have seen, move unstably between the four institutional styles, and therefore between their styles of judgment. Therefore, all that one can say in general about the kinds of e-governance tools that are most likely to serve their purposes and prior commitments, are that they will need an awareness of the range of tools that they could use, and should look for those tools that most fit the institutional style or combinations between styles that they think of themselves as occupying.

It is in this area that there is the greatest need for experimentation to find out more about what kinds of disciplines work with different groups of policy makers at national, local and transnational levels, to sustain judgment as a settlement between the four institutional styles.

The challenge for policy makers: policy making, constitution and e-governance

Using e-governance technologies to support conciliation between institutional styles

The argument of Chapters 8 and 9 was that we ought not to rest contented with finding new tools with which to further the existing commitments of the four basic institutional styles, but that there are good reasons for wanting, not to eliminate any of these situations for all are important, but to develop better ways of conciliating between policy makers in these different situations, whilst accepting the fragility of four-way settlements. For the process by which policy gets made is most likely to be reduced to gridlock or irreconcilable conflict in those contexts where the institutional styles become polarised, and where people in each of the four situations move to the furthest edges of the matrix used in Chapters 8 and 9, when the dimensions of constraints and bonds conceived are differentiable[17] and not only as oppositions between four stable 'attractors' (Thompson, 1996; Eve *et al.*, 1997 and Byrne, 1998).

If the argument of this book is accepted, that policy making is ultimately about sense-making long before it is about making particular decisions, then the challenge for conciliation and settlement between institutional styles must itself be a process that facilitates some convergence, however modest, between the rival interpretations produced in alternative institutions which collectively make up the activity of politics (Crick, 1964 [1962] and Gamble, 2000). Conciliating in sense-making involves efforts to create contexts for thoughtful joint deliberation, moderate the polarisation of identities around institutional situations and styles, finding points of convergence in what appear to be irreconcilable versions of history, creating a sense from the political projects that emerge from different institutional settings, and experimenting with alternatives (Weick's 1995, 2001, ch. 21). By such means alone can political judgment be sustained and cultivated, understood in the round as I argued in Chapter 7 that it should be.

The aim cannot be to hope for convergence. Conflict of perspectives and interests within the system of governance, as well as in the wider world of politics, is not something regrettable, save when it has become polarised. On the contrary, it is necessary to viable governance, and most especially to democratic governance. The systems theorist's 'requisite variety' theorem that any organisation of governance must have enough variety within it to reflect the problems which it must address (Ashby, 1947) is not merely a piece of sound technician's folklore, but a normatively important principle of design in policy making. The challenge is to limit the polarisation of that internal conflict, not to eliminate it.

The reasons for pursuing such settlements are ones that, finally, have little or nothing to do with technology, and everything to do with politics. But there is no reason why e-governance technologies cannot be used in the course of developing institutional settlements between the basic styles that are thrown up by the situations created in any policy-making process. In this context, both the formal models of the policy process and expert systems representing rigid structures of rules, and also the individual use of tools are of less use than are the systems that support some form, however thin and opportunistic, of communication and collaborative deliberation about preferences, and which are open reasonably wide to participation by people in each of the settings in the policy process. For example, one can readily imagine the use of a series of electronic conferencing, argumentation support, organisational memory capture tools made accessible through extranet sites which can be used by policy makers across the legislature and executive, in order to sustain multilateral dialogue, being useful in the course of initiatives to this end. Such programmes could readily integrate access to problem-structuring tools, facilitate the use of intelligent search agents that individual policy makers could personalise, and could also offer links to formal models that can be interrogated online in order to run rival assumptions and display results, albeit more as illustrations and ammunition in argument than as definitive exercises in settling policy questions. On an advisory rather than a binding basis, online conflict modelling and management tools might also be used in support of particular efforts in conciliation between policy makers in different settings and with rival commitments.

However, if the argument in Chapters 8 and 9 is accepted, what is really important in using e-governance technologies in any of these ways to support political judgment as a form of settlement between institutional styles, is that their use is embedded in ritual-like forms or disciplines of sense-making and, second, of decision making, that to some degree gives recognition to the place of each of the four basic institutional styles. There are, as we have seen, more and less well articulated ritual forms, procedures and disciplines, through which, within each of the institutional styles, policy makers engage in sense-making, learning, decision making and generally organise time in the process of judgment. What are less firmly institutionalised, however, are the ritual forms that support settlements between the institutional styles in which the contribution of each is recognised. There are, of course, stylised ways in which issues flow through the hands of policy makers in each of these types of settings: these provide sequencing or minimal policy coordination, but not much administrative integration (6 *et al.*, 2002). For example, issues are initiated by movements and factions, finally and reluctantly put onto the agendas of the competitive leaders, delegated to central professional units, returned to the leadership who at this stage may face the problem of helping the more isolated backbenchers to deal with the backflow, and finally are remitted to the bureaucracy for implementation, leaving front line staff who are often in either semi-isolate and semi-movement situations to shape them as matters of

administrative discretion. There are also many ritual-like ways of organising the relationships between the settings which simply subordinate
all contexts to a single one, as in the common case in which competitive
leaders simply, *de haut en bas*, invite comments upon proposals of their
own. In a system such as that which characterises the British case, in
which primacy of politics and the sovereignty of the legislature largely
mean the dominance of a cadre of competitive leaders in the ministerial
and very senior civil service tiers of the executive,[18] this hegemonic
ordering of one institutional style is only to be expected. The advantages
of such an ordering are well known and need not be denied: there is a
democratic case to be made for such a system, there is some evidence for
its efficiency, and it has the merit of comprehensibility to the public and
to the policy and political élites. However, this hegemonic ordering also
has its costs. It can create bottlenecks in the flow between institutional
settings of information and, more importantly, of alternative interpretations that are essential to robust sense-making, and it can lead to a
reduced ability or willingness of people in the subaltern settings in the
policy system to 'speak truth to power' (Wildavsky, 1980).

In other countries with other traditions, of course, it is sometimes
easier to develop a wider range of institutional arrangements. The
constitutional order in, for example, the Netherlands allows for much
greater integration between the institutional settings without overarching hegemony for the cadre of competitive political leaders and a
mandarinate of the most senior civil service. Federal constitutions such
as those of the United States, Canada or Australia introduce other
flexibilities into the ways in which settings can be related to one
another. Federal systems that give representatives of subordinate geo-
graphicals a place in the upper chamber of the central legislature, as
does Germany's basic law, provide a measure of integration on at least
that dimension, although integration across departments of the executive remains weak in German federal governance.

One key limitation in the British system's capability for enabling
greater integration without hegemony for the cadre of competitive leaders, is the erosion of a clear and well institutionalised constitutional distinction between state and government. As recent administrative law
decisions in judicial reviews have increased the accountability of the
civil service to the government of the day at the expense of independent
accountability to the 'constitutional fiction' of the Crown in
Parliament, as the constitutional processes that stress the Crown rather
than government have come to be regarded as part of the merely 'dignified' part of the constitution in Bagehot's unfortunate phrase, and as

local government has come to be firmly subordinated to the centre in both finances and policy goals, the space has been reduced for non-hegemonic and more flexible ways to relate institutional settings of policy making to each other.

The central long-term challenge, then, for a robust policy process, and for the development of the use of e-governance tools within it, is to develop a plurality of ways in which the institutional settings that make up the policy process can each make their contribution effectively.

The pivotal importance of the legislature

Where can we look for institutions that could provide some conversation over sense-making between policy makers in all of the four basic institutional settings, that could at least in principle support the disciplined and ritual-like practice through which conciliation over sense-making might be conducted? Essentially, there are two kinds of such institutions, namely, those which are essentially membership structures, and those which are practices or ways of doing things. In most democratic countries, the principal membership structure that brings together the competitive leadership cadre, the central professional units, the isolates and the factions is the legislature.

Providing adequate e-governance tools to the legislature has traditionally been conceived as a matter of providing electronic voting systems for voting on bills in committee stage or on the floor of the House. However, the present argument leads clearly to the conclusion that it is time to think about the issue much more broadly, and not only in terms of particular technologies, but in terms of the kinds of political function we expect of the legislature in supporting political judgment as a practice of conflict containment in sense-making between differently situated policy makers.

The scrutiny role of the legislature is of crucial importance, by which the executive comprising the competitive leadership cadre and the central professional units as well as the implementation agencies, are held to account by a body comprising isolate backbenchers, other rival bodies of internally competitive leadership groups in opposition parties, and several movements or factions. When it becomes subordinated to the other role of the legislature, in supporting and sustaining the executive, the very dangers emerge that have worried both liberals and democrats for centuries. Indeed, although one focus of that liberal and democratic concern was always upon the risk that an executive inadequately held to account by the legislature might abuse its power, another was that the process and the quality of political judgment would become less robust by the erosion of the capability and willingness of those outside the executive to 'speak truth to power' and engage in a conversation with the executive on terms that, although never quite equal, nevertheless give clear *locus standi* to the interlocutors.[19]

At present, in the British case, the main instrument of legislature scrutiny is the Select Committee system. As was argued in Chapters 8 and 9, scrutiny committees can exhibit any of the four institutional styles and can combine or strike settlements between institutional styles, which makes them a particularly important resource for the larger goal of sustaining political judgment in the political system as a whole. Unfortunately, it is old news that in the British House of Commons, only the Public Accounts Committee is taken seriously by the executive, and that few other Select Committees are able to secure the access their most active and independent-minded members would like to the senior civil servants they would like to hear from, or indeed even to ministers outside the particular department they are principally charged with examining, but whose routine work or special initiatives may nevertheless greatly impact upon their field.

There are a number of things that could be done to improve this situation, and to which the use of new e-governance tools could be highly relevant. First, there is the question of what executive information Select Committees should have access to, as of right. Today, in the United Kingdom, they are largely dependent on information that they can glean by face to face interrogation, and by the individual members using the system of posing written Parliamentary Questions. With the exception of those paper documents that are not generally published but are placed in the House of Commons library for any Member of Parliament (MP) to view, members of Select Committees have only the same access to information created and used by the executive that the ordinary public has. That is to say, individual members who wish to, can view information published generally, either in paper form or else on government websites. The only committee that has access to otherwise privileged information, and that subject to confidentiality rules, is the committee charged with oversight of the security and intelligence services.

There is a case for creating a special category of authorisation and security, perhaps – depending on the level – subject to certain duties of confidentiality for certain categories of information – of access to the Government Secure Intranet, for Select Committee members. There are many documents and analyses that are currently available through the GSI that it would greatly assist Select Committees to have access to, and where there are few, if any, major problems in their having a right to see them, either in the form of perverse incentives for executive politicians and officers not to write things down at all, for in the form of possible compromise of the independence of the legislature from the executive. True, there are obvious reasons why they should not have access to every document or data set. There are always confidential pieces of advice to ministers that will remain privileged (at least unless or until they are released under the thirty or the fifty year rules), for which the problem of perverse incentives arising from wider access is serious. However, there are many data sets in particular, and many formal models to which Select Committee members ought to have access, and by the ability to conduct analyses for themselves on those data sets and models, their scrutiny capabilities would be greatly enhanced without compromising the position of the executive. More generally, there are documents created within the executive that are not particularly sensitive, when considered in the context of the possible media profile, but which reveal the range of ideas and debates within the executive about policy, that could and should be available to Select Committee members, but may not be, principally because MPs simply do not know of their existence and therefore are unable to ask for them, even if they would not be denied them in paper form, were they to ask.

Second, the example of the Finnish Parliament's Committee for the Future shows that there is real value for considered political judgment for legislators to be able to use e-governance tools to gain access to experts for conversation and interrogation in real time, using such e-governance tools as videoconferencing. The current practical and financial restriction of Select Committees to face-to-face interrogation of ministers and senior civil servants is closely linked with the effective restriction on the numbers of people they may call, and with their limited constitutional ability to get below the Permanent Secretary level to reach policy staff whose work is of real importance. There is, therefore a case that both the constitutional restriction and the practical limitations could and should be relaxed together, enabling committees to conduct interviews with senior civil servants (down to some specified level, defined either by grade or by function) whom they identify, and using videoconferencing or other online means including perhaps the same technologies used to support 'chat rooms' for real time synchronous conversation on websites. This would both expand the reach of scrutiny and at the same time reduce the time demands that face-to-face appearances before Select Committees can make upon civil servants. Of course, if Select Committees abused such new freedoms by unreasonably demanding too much time of certain staff, then additional safeguards might be needed, but it would not be difficult to devise such safeguards or to explain their importance to the House.

Third, Select Committees ought to have greater access than they do today, with the present limited budgetary support for their work through the British House of Commons' own budget, to those e-governance tools and systems that they consider of real importance to their work. There could be, for example, a fund created, to which Select Committees might apply, and which might be controlled by the Liaison Committee of the House which has responsibility for the Select Committee system, which would provide financial support for Committees that can make a case that their work of oversight and scrutiny would be greatly helped by having access to new tools. The kind of support provided by the leading business school to the all-party that was described in Box 3-b in Chapter 3 is a good example of the kind of e-governance tool that might be of use to many Select Committees as they deliberate upon candidate recommendations for their reports. Again, access to support in the use of formal models, or resources to conduct online public consultations, where a Select Committee is charged with examining a draft bill in a year prior to its expected inclusion in the legislative programme, would be other examples of the kinds of use of e-governance tools that could usefully be supported by such a fund.

Fourth, there is a case for creating a similar fund, although perhaps a smaller one, to which all-party groups might apply. All-party groups are an often underrated aspect of the scrutiny function of the legislature, yet many of them do provide essential information and invaluable fora for deliberative processes among legislators. Their constitutional status is shadowy at best. Without going so far as to give them any entrenched constitutional rights, there is no reason why the House should not create a modest fund to encourage their work by supporting, again on a competitive basis, their use of new e-governance tools.

Finally, there ought to be some responsibility within the legislature for taking a regular overview of the development in the use of e-governance tools within the House. The logical place, presumably, for that responsibility to lie is with the Liaison Committee, which comprises the chairs of all the Select Committees and which has the responsibility of ensuring the health and authority of the Select Committee system. That periodic review should attempt to draw out the lessons from the initiatives taken by particular Select Committees, and should address the question of which technologies appear best to suit scrutiny bodies characterised by different institutional styles.

The executive

Within the executive, most of the debate about how new digital technologies can support joined-up or holistic government has been concerned with the design of systems for service provision. However, if we have learned anything of real importance from experiments in holistic electronic public services in recent years, it is that the blithe promise of too many information technology companies must be set aside, that integration at the customer interface can entirely substitute for sorting out the inter-organisational relationships and for the need for working on coherence across departments and agencies on policy.

The commitment to institutionalising holistic working among policy makers requires more, if the argument of this book is accepted, than merely someone checking the literal consistency of proposed laws and particular decisions coming from one department with those of another. It requires work to support collaboration or at least some conciliation in the ways in which policy makers make sense of their problems and their experience, and in the ways in which they come to judgment. For this purpose, the archival approach adopted in many uses of , for example, Britain's Government Secure Intranet (GSI), to enter documents into banks of policy materials that can be picked up or not by other departments, is nowhere near enough. It also requires more than the use of e-mail to circulate draft policy documents to people in other departments. These are only incremental changes to existing practices. Whitehall has long been very good at bi-lateral, rather passive and separately deliberative policy-making processes, and these already have their established ritual-like forms. The next stage must be the development and institutionalisation of more holistic policy making in more multilateral, more active and more jointly deliberative processes, and ones that reach beyond the 'power diagonal' that characterises the British system, between the cadre of competitive leaders and the central professional policy analysis staffs.

If policy makers are willing and able to innovate in these ways, then there are, as we saw in Chapter 3, a range of e-governance tools that would support those innovations, but, as we saw in Chapters 4 and 5, purchase of the technologies alone will not institutionalise organisational innovation.

Although centralisation is not necessarily the principally or typically most appropriate strategy in response to problems of fragmentation in policy making, there is an important role for the centre in providing a kick-start for more decentralised processes of coordination and integration (6 *et al.*, 1999, 2002; Bouckaert *et al.*, 2000). Unfortunately, such central funds as are available for supporting e-government initiatives have been conceived almost exclusively as in the context of the promotion of innovation in e-services (Performance and Innovation Unit, 2000c). It would be helpful to use these funds for bids to support new practices in collaborative policy making that require particular e-governance technologies to support them. However, it should be made clear that departments and agencies working together should not be bidding just for hardware and software, but should be developing new ways to make policy, to exercise judgment and to engage in collaborative sensemaking, focusing principally upon the disciplines and the techniques that fit their particular problems, but for which particular e-governance tools are necessary. Second, the executive centre also needs to create the capacity for regular review of what lessons can be learned from initiatives and experiments in policy making and the use of e-governance tools to support it.

Local government

Having made out the argument for changes to both the elected and the salaried tier of policy making to support processes of conciliation in political judgment at the level of central government, it is possible to be quite brief in dealing with ways forward for local government.

Straightforwardly, the same principles should apply to local scrutiny committees that, I have argued, should apply to Congressional Committees or British parliamentary Select Committees. Scrutiny committee members should be granted online access, to appropriate levels, to materials on local authorities' intranets, should have the right to call officers below the level of chief officer, should be able to draw upon resources to support their use of e-governance tools, and the lessons from experiments in scrutiny committees' use of e-governance tools should be drawn together both within authorities and nationally. There is no particular merit in creating another special discretionary fund for this to be administered by the central department that has responsibility for local government finance. If resources for this cannot be built into the centrally set formulae, which would be preferable because this would allow authorities to make their own decisions about how to develop scrutiny e-governance, then it would probably be more appropriate to give the responsibility for administering a central fund to the collective representative body for local government.

Again, the same principles that were applied to the case of executive policy making in Whitehall should apply to local authority executive policy making, and, likewise, if for political reasons, increments to central support cannot be made for the purpose of supporting developments here, then it would probably be more sensible to devolve responsibility for administering any central discretionary fund to the representative body.

The challenge for the information technology industry

If, as was argued in Chapter 4, we can expect greater pressure for the use of e-governance tools, then the information technology industry will also need to give greater thought to what it can offer policy makers. Policy makers represent a potentially large market, including the several tens of thousands at the centre, similar or even greater numbers in local government, the National Health Service and the plethora of agencies and non-departmental public bodies.

The first and, on the evidence of the industry's past record, greatest challenge, is to develop tools that support sense-making rather than simply focus on supporting decision making (Weick and Meader, 1993), and to market them in ways that reflect this changed understanding. The industry has too long imagined that its main clients either are exclusively within the central professional units, or else can be made to work with information in the same way as central professional policy analysis staff do. The argument of this book is that the contrary is the case. Designing and marketing on the assumption that increasing sophistication in the technocratic features of policy design as optimisation, or that simply access to more and more information, are what is required, were never viable strategies. The industry needs to learn the lessons from the history of slow take-up of these tools. The information technologists who design e-governance tools will need to get much closer to those working on the disciplines and ritual forms of sense-making than they have done so far. The industry needs to understand the very different requirements, information games, styles of judgment and styles of learning exhibited by policy makers in the different institutional settings, and both design and market e-governance tools accordingly.

Second, the argument of this book that what is really important in the long term is finding ways to open up conversations between differently situated policy makers, brings to the fore the issues of systems integration, not only across executive departments at central and local levels and between them, but between the executive, the legislature, local elected councillors and local executives, and the plethora of regulatory agencies with responsibility for oversight of the executive from the Environment Agency to the Information Commissioner.

E-governance and liberal democratic renewal

Perhaps by this stage in the argument, some readers may have qualms about the way in which judgment and policy making have been discussed, as principally activities conducted within a charmed but hardly charming circle of élites, including politicians and mandarins, local government chiefs and agency staff, professional policy analysts, regulators and such outside experts as they choose to draw upon. Surely, some readers will protest, a democratic political culture demands more than this? If e-governance tools merely enhance the capabilities of the élites to exercise judgment, but do nothing either to open up the process of judgment to popular scrutiny or even to greater democratic participation, if they have no structured relationship with the ambitions for the use of digital technologies for renewal in democratic life, then, it will be asked, are they not positively dangerous?

This is, of course, by no means a problem that is new or that is specific to digital technologies. It not only could have been, but was raised in connection with every previous technology used to support governance. Greater democratic control over the collection of statistics about populations was an important strand in democratic movements in the nineteenth century in response to the development of statistical surveillance as a tool to support policy and political judgment from the mid-eighteenth century, and it has been shown in popular concern about the census ever since.

The challenger is quite correct to point out that it is not enough for me to reply, lamely but quite truthfully, that there is no inconsistency between a programme for the wider use within the political and policy élites of tools of e-governance and a programme for greater popular participation in the polity, whether by using digital technologies or any other means.

An adequately detailed answer to the challenge cannot, for reasons of space, be developed here. However, such an answer must consist in at least the following four strands.

The first strand is concerned with the ways in which e-governance tools can and should be developed as part of reforms to the policy process that, at the very least, enhance democratic accountability through the greater openness and accountability of the representative institutions. The argument that has been presented in this chapter about how these tools can be used to support an enhanced scrutiny role for elected representatives is a central element in this strand of the answer to the democratic challenge.

As noted above, one way in which e-governance tools can be used to enhance the accountability of the representative institutions concerns the use of filter systems by which US senators and now some MPs are beginning to cope with the volume of incoming e-mail. Many are installing e-mail filter systems that parse sentences in each message, and sort messages into those sent in one of several pro-forma automatic replies, those forwarded to someone else and acknowledged with a note about forwarding, those to be dealt with in person by junior staff in the legislator's office, and finally those to be dealt in person by the legislator, and so on. As a way to cope with information overload without hiring ever more administrative staff, the use of these tools is reasonable. It is reasonable to ask that legislators should publish on their websites, the criteria by which these systems are instructed to classify and filter different kinds of communication, so that these criteria can themselves be the subject of comment and public debate among constituents.

Also, if the recommendations made above are followed, e-governance tools can clearly be used to support greater responsibility and autonomy from the centre, of local authorities to take greater roles in policy making and not simply in acting as agents for the implementation of centrally set policies and this greater local governance is a basic demand of many democrats concerned about the representative institutions.

A second strand of the answer is concerned with the importance of integration between systems of online consultation with consumers of public services, non-consuming taxpayers, and wider publics on the one hand, and systems to support sense-making and judgment within the policy élites on the other. There are powerful arguments for much greater organisational and information systems integration here.

A third strand is concerned with the ways in which e-governance tools can be used to support freedom of information and more openness at the level of the executive. That there are limits to what can be published at the time when it is being developed in the executive is well known and well understood. However, there are many formal models, many data sets including GISs and other tools that can be made publicly accessible over digital networks, and to do so would significantly enhance the democratic process.

Some of the limitations upon the openness of government are important in the context of the debate about e-governance. In particular, the recommendation by the British government's Performance and Innovation Unit (2000a) that all models should be made publicly available is, in principle, sound, but is appropriate mainly in respect of models of policy problems. Where policy makers use e-governance tools to develop models of their own provisional preferences which may well be revised before final and formal taking of decisions, as is done in chauffeured decision conferencing (Box 3-e in Chapter 3 above), there are often good reasons for wanting to protect these from publication, for otherwise policy makers with tendencies to poker-playing will be even more inhibited from candid and collaborative exploration than they already are.

To the extent that e-governance tools are used to support more holistic working in government, and to the extent that holistic working is designed to promote outcomes that the public care most about (6, 1997a), there are powerful democratic defences to be made for their use.

However, there is a fourth strand to any such answer which is much more controversial, and even dangerous to discuss, but which is of enduring importance in a liberal, as opposed to a mass or simple majoritarian or plebiscitary democracy. This book has argued that political judgment is of supreme importance, that it is not something that is 'possessed' or 'achieved' by individuals, still less by political castes or classes, but is produced through the process of politics, in the course of the interaction between people in different institutional settings. The policy-making élites have no monopoly upon good political judgment, and their accountability to the popular will, fragmented and conflict-ridden as it too must be, is crucial to their capability for achieving judgment. But neither is it the case that the popular will or public opinion 'possesses' judgment, or that the shared sense made by the current majority must always and everywhere trump the shared sense made by the élites. As the advocates of the US constitution have always argued, the judgment process is something that should be so designed as to appeal to the highest and the best, long-term, underlying, reflective and deliberated upon commitments of citizens, not necessarily to their majority's mood of the moment. Some things are put, constitutionally, beyond the reach of temporary majorities for just this reason.

But the argument cuts deeper still. A decent and liberal democracy is one in which the contribution of popular sentiment to the process of judgment is institutionally produced, acculturated and sustained very deliberately. It was precisely because of this recognition that in Britain, after winning the vote for women, may suffragettes went on to found the Townswomen's Guild, with the aim of educating women to use their vote and to use it with deliberation, whatever their considered political views. It has been from this very argument that the justification has come for the range of recent initiatives for citizenship education in schools. For a citizenry has no means of cultivating and exercising the vigilance over its most fundamental concerns, that is its own responsibility in the accountability process, if the conception of democracy is simply one in which people express their current preferences and majorities are calculated.

A proper conception of e-democracy, and of its relationship with e-governance, should reflect the same recognition. The measurement and translation into action of the current mood of popular majorities is not an adequate aspiration for a democratic process that seeks to promote judgment. One of the peculiar responsibilities of policy élites is to use their own processes of deliberation and judgment to provide the materials on the basis of which citizens can deliberate and, in some cases, stand back from their immediate passions, and in others, accept that having temporarily and subject to the strictest conditions of accountability at the ballot box, through the media and by many other means, delegated a measure of authority to a representative government, that government may have both the right and the duty to act upon considered and deliberated judgments that may be unpopular in the short term, that may have to impose costs that will be resented, but that doing so may nevertheless be the least worst or even the right thing to do. E-governance tools may not merely assist policy élites to make those judgments, but also to explain and legitimate them publicly, and to give account of their actions, which is the first and most basic element of accountability (March and Olsen, 1995, ch. 6).

This is certainly not to argue that political and policy élites have any fundamental claim, on the basis of who they are or what they know or the power they wield, to a one-way tutelary role over the citizenry. The rejection, or at least the heavy qualification and restriction, of paternalism and the paedagogical or tutelary state goes to the heart of both liberal and democratic thought (6, 2000). However, the institutional process by which political judgment is achieved is a dialogical one, in which every group both within government and among the wider publics, is both teacher and learner. The only intelligent way in which, in a liberal democracy, one can respond to the eternal debate between Paine's insistence on popular sovereignty and Burke's contrary insistence on the duty of the delegate to exercise judgment, is to recognise that both are equally necessary.[20] Basic political decency requires that this reply to the democratic challenger be made with some element of regret, and certainly with no relish whatsoever, and having first made clear the importance of all the safeguards of accountability to the popular will, but it also requires that it be made firmly.

It is here that the really important dimensions of the relationship between e-governance and e-democracy become clear. The first three strands of the reply to the democratic challenger are of central and fundamental importance. But, the fourth strand makes clear, the challenger is absolutely right to detect that at some deep level, there must be a tension between e-governance and e-democracy. Tensions need not lead to catastrophic polarisation: when they do not, they are creative, vital and profoundly valuable. The point of the fourth strand of the reply is to say that the key challenge for institutional design is to ensure that this tension is of that kind.

In a democratic polity, policy making *should* reflect a wide range of types of information that are counted by each of the institutional styles as policy relevant information, and not only technical evidence about the cost-effectiveness of interventions. Popular support, acceptability with key constituencies, conformity with constitutional norms sometimes argue for policies that may not turn out to be particularly cost-effective, but the point of a democracy is to give these factors due weight in the minds of policy makers. Better policy is more likely to come from the frank acknowledgment that in a democracy, we positively want a system that gives recognition to each kind of evidence and judgment, to each kind of interest, and not just to those who control the slide rule, but also from those who mobilise their allies, those who control their own vote, and even those millions who simply want to get on with their lives as best as they can. Achieving this requires a politics that acknowledges the place of all four institutional styles, not only among the élites but among the wider population of citizens as well.

11
Conclusion: Technology, Institutions and Political Judgment

In the early 1990s, the economist Robert Solow once famously remarked that one could see the effects of the computer revolution everywhere, save in the productivity statistics. At long last, it does now seem to be showing up there too. However, to adapt his phrase to the present context, by the mid-1990s, one might have said that we have seen the computer revolution everywhere in governance, except in the quality of political judgment. As Chapter 3 showed, this is no longer entirely fair. The effects have been, as Chapter 4 explained, a long time in coming, but finally there are signs of innovation. A number of pressures are building for their greater use in policy making.

However, a central argument of this book has been that neither those technologies nor the techniques of their use will in and of themselves produce or enhance political judgment. Those whose ambition for the information technologies remains that of the early pioneers of the arti-ficial intelligence movement – namely, the wholesale automation of thought, the substitution of rigid algorithms for complex balancing in the context of dialogue, negotiation and shared deliberation, and the elimination of politics from policy – will surely be disappointed.[21]

On the contrary, people use information and knowledge technologies for the political purposes they already have under the constraints they already work under, but those purposes and the conditions under which they can use those technologies are not ones of their choosing, at least in the short run. That is the work of institutions, and often of a rather informal kind of institutions, rather than of the formal constitutional rules upon which political scientists typically focus. However, most of those who have focused recently on institutions in policy making have tended to write as though those institutions could only be in one of three conditions – namely, either harmonious and promoting good policy, or else in conflict and leading to bad policy and in need of reform that will align those institutions with each other, or else that institutions will simply create pressures for empty pretence and systematic hypocrisy ('symbolic politics'). An older generation of thinkers about politics were wiser, in recognising that conflict between institutions is at once necessary, unavoidable, and, within certain always ill-defined bounds, creative and valuable in sustaining practices of judgment (e.g. Crick, 1964 [1962]).

Whether or not the generation of e-governance technologies that has become commercially available in the last ten or more years, classified and discussed with examples in Chapter 3, make a positive contribution to the quality of policy making and of political judgment depends principally on the choices we make about the way in which policy makers in different situations and groups in structurally similar settings in the wider society try to relate to each other. The settlements we can strike between institutional pressures under which policy makers try to use these technologies will be crucial.

Therefore, it is probably a mistake to ask, as indeed I myself did when I began work on this book, 'are some countries – the United States, perhaps – leaders, and others – Britain, for example – laggards in e-governance?'. To be in the forefront of the adoption of particular technologies to support policy making for its own sake, and without trying to innovate in the institutional process of conciliation between rival institutional settings and styles of political judgment, is not an ambition worth pursuing, and a crude league table of adoption of hardware and software is not a league table worth compiling.

Political judgment is centrally about sense-making, long before it is about decision making, and the most urgent need, in any programme for the improvement of the policy process to support judgment, is to support policy makers' capabilities for making sense of a fast-moving, opaque, equivocal world (March, 1994, ch. 8; Weick, 1995, 2001), in which their understanding of the impact their policies and actions are having will always be limited, and will always reflect in part the institutional setting in which they must do that sense-making. Our best hope for the crafting of institutions within which technologies of e-governance can be best used effectively and judiciously, is to look for ways of bringing those rival institutional settings into conversation with one another, without ever wanting or expecting convergence. The real threat to the calibre of judgment arises from their polarisation. But with greater institutionally buttressed conversation between policy makers in different settings, even if that conversation can – contrary to the high hopes of the thinkers of the Enlightenment age – sometimes only be conducted at all on the basis of mutual misunderstanding, we have some chance of finding ways to help policy makers make sense and come to judgment. So far from privileging them above the citizenry, in such a context, e-governance tools could be used to sustain precisely the important elements of accountability and tension that a liberal democratic order requires. That is why e-governance is much the most important issue in all the debates about electronic government, and also why it is of vital importance that it be more widely and publicly debated.

Notes

1 Introduction

1. For a comparison between this usage and the current rage for using the word to describe joint activity between public, private and non-profit organisation, see Pierre and Peters, 2000, chs. 1–2. For a statement of the currently more fashionable usage to mean little more than the exercise of any kind of power by the interaction of actors and organisations in every sector, see Kooiman, 2003.

3 Examples and Initiatives

2. Although the number of appropriate frames for a given problem is not necessarily bounded, fairly obviously, some frames are simply inappropriate to almost any plausible policy-related interest at least within the executive, and this fact represents an important constraint on the ways in which we think about frames: Goodman, 1978; Elgin, 1996.
3. Milner 2000, 113–16, citing White *et al.*, 1998, argues that projects in knowledge management typically go through a series of stages of expanding ambition, disappointment, refocusing and new investment, and eventually cease to be seen or managed as stand-alone technology initiatives at all, and become part of the general strategic direction of the organisation.

4 The Forward March of E-governance Halted?

4. For an argument that with careful selection of technologies to fit to the culture of the organisation, with the right champions at senior level to elicit and lead commitment, and with proper training for senior executives, such systems have great potential, see McNurlin and Sprague, 1998 [1993]: implicit in such arguments is the view that low take-up reflects the failure of the human side of organisations and their consultants and suppliers to supply these prerequisites. For an argument that the fault lies with the design of the software designers, see Weick and Meader, 1993.

5 What Difference could E-governance Make?
Fears, Aspirations and Theories

5. This is not to underplay the importance of economic, demographic and other material factors in shaping organisations, but it is to give appropriate stress to the importance of the way in which organisational – and indeed political – forces mediate the work of these factors upon governance. For a major account of information technological change that gives due weight to economic, demographic and material factors, but which underplays the partly independent role of organisational dynamics, see Castells, 1996.

7   Policy Making: the Work of Governance

6. There is an extensive debate in the literature on political judgment flowing from the writings of Hannah Arendt, concerned with the question of how far political judgment can or should be independently political, and how far it can and should be fundamentally the application of moral judgment to political particulars: see Beiner and Nedelsky, 2001 for a recent collection of articles, with Kateb (2001) there arguing for the moralistic position, and objecting to the idea of an autonomously political character for political judgment as aestheticised politics, which, in his view, runs close to the totalitarianism that Arendt combated so courageously. Without entering into the exegetical debate about what Arendt meant – which is in any case resoluble, since Arendt did not live to write the third volume of her *Life of the mind*, which would have been her systematic treatise on judgment – it is important to stress that astute political judgment cannot but be activity of sense-making, and narration. As Kateb acknowledges, Arendt was sensitive to the importance of storytelling, not only in the work of leaders – as more recently, Gardner (1997) has been – but in the judgment of ordinary people. Part of Eichmann's incapacity for judgment, in Arendt's argument, was indeed his utterly diminished moral capacity, but another crucial part was his inability to exercise the facility for sense-making. His lack of what she would later come to call 'enlarged mentality' was his stunted *political* imagination, that he apparently could not conceive his experience of his victims in rounded narratives, but only as cases to which he applied the bureaucratic form of the rules of *Führerrecht*. An Eichmann with that capacity but without the moral capacity would perhaps genuinely have been a monster, and the evil of such a person would not have been merely 'banal'. On the logic of Arendt's argument, such a person would have possessed one component of political judgment. Perhaps that is why we tend to regard Himmler and Goebbels as monsters, and not merely as people performing banal, bureaucratic evil. If this is the correct reading of Arendt, then it seems to me importantly right. To summarise, political judgment involves sense-making; astuteness in political judgment involves the balancing of rival forms of sense-making, including importantly the moral, but without necessarily, automatically and in advance, privileging any particular style of moral sense-making. Political judgment is certainly not immoral, nor is it amoral, for it cannot but take cognisance and fore-sight of the likely future accountabilities of plural moralities, but it is much more than the simple application of a scheme of moral principles. A morally good and virtuous person must not be adroit, but in addition, they may, after all, not be astute, in the sense that I use the word, either. Therein lies, as I understand it, one of the most crucial distinguishing features of political from moral judgment.

8   The Organisation of Policy Makers and the Use of E-governance Tools

7. This is not to say that the four forms to be distinguished are *only* ideal types, in the sense of having only the status of heuristic devices. Indeed, on the contrary, it can be argued that their significance is much greater than this, and

some reasons for thinking so will be given. However, the nature of the present enquiry does not provide sufficient evidence to support the larger claim in full, and therefore the theory can only be used in this book in the manner of ideal typical explanation.

8. By 'style' here, I mean a distinctive *discipline*, or a body of socially biased practices with a certain social authority (Durkheim, 1961 [1925a]: this is not the same as Foucault's rather extreme usage of the term) or an institutionally ordered and internally consistent form ('stylisation') of social relationships and of gathering, using, interpreting and trust others with information: for the origins of the concept of style here, see Goodman, 1978, Ch. II; Mannheim, 1985 [1936]; Douglas, 1996. For a source discussion of the institutional dynamics of stylisation as the exaggeration of the core commitments of institutional forms, see Durkheim, 1980 [1906]. In that review, Durkheim writes: 'A political system, once initiated, sets up a system of ideas on which it relies and which justifies its rationale. Then this system of ideas, in its turn, once it is carried through, reacts upon the political organisation to which it first simply gave expression, and alters it. Generally, it tends to exaggerate it. For the standards, the formulas, that are invented to give it a logical foundation are generally, much simpler, more exclusive, than the historical causes that brought about its genesis and affected its spontaneous growth.'

9. The two dimensions can, of course, also be read as continuously differentiable, rather than as binary and discontinuous. For a continuist reading, see Gross and Rayner, 1985, and for a discontinuist reading, see Schwarz and Thompson, 1990 and Thompson *et al.*, 1990. In fact, both readings are illuminating. The four syndromes are distinct and at least provisionally stable, and do, as we have seen in the previous section, describe some clear cut cases very well. These four stable styles seem to have a distinct power and appeal, functioning as 'attractors' if the matrix is read as the kind of space described in complexity theory: see e.g. Eve, *et al.*, 1997; Byrne, 1998. In essence, then, the four basic styles are 'mechanical solidarities', as Durkheim, 1984 [1893a], defined them, or institutional styles of social organisation. However, there are surely degrees of strength and weakness, each with varying degrees of stability. As we shall see when we come to consider mixed cases and settlements between the basic four institutional styles, such mixes and settlements are easier with more moderate forms of the basic styles. For a discussion and an attempt to offer an interpretation that preserves some of the virtues of both, see 6, 2003, esp. p. 14 note 3.

10. I stress here the causal nature of the connection, because in some formulations using this style of explanation, it can appear as though the connection between these forces and the characteristics of each situation is postulated in the definition of each position, and this is specifically not what I wish to argue.

9 Institutional Styles and E-governance Tools

11. The point here is that each of the institutional styles leads to an exclusive focus. The phrase is from one of the late Sir Isaiah Berlin's most famous essays, quoting a maxim from the ancient Greek poet, Archilochus, that

'the fox knows many things, but the hedgehog knows one big thing': Berlin, 1978 [1951], 22.

12. Douglas, 1966a, ch. 4 shows that some things cannot be experienced without ritual. In particular, the orientation of time which is at the heart of the account of political judgment offered here is something that, Douglas has shown, is achieved principally through ritual. Douglas and Isherwood, 1979 show that ritual is essential in order to 'fix public meanings', which is a crucial element in sense-making. Westrup, 1996 shows that rituals are used to fix the roles of participants in meetings, to reduce uncertainty, to increase trust – at least within the institutional style of the organisation he was studying, which was not composed exclusively of competitive individualists or isolates – and, crucially, to enable people to make sense of their experience, by giving structure to the time over which a consultation process was conducted, and giving a narrative structure to the exercise of judgment in which they were engaged.

13. Bell, 1997, ch. 5 argues for a very tight definition of ritual as confined to certain genres, none of which are of much significance within the policy process, but allows a variety of activities to be more or less 'ritual-like' to the extent that they involve at least some of the features of formal discourse and procedure, the production and reproduction of traditions [within organisations], are more or less invariant, rule-governed, use symbols to which some status is attached, and are conducted through something like performance. In this sense, many of the ways in which policy makers conduct meetings within their organisations and between organisations in different institutional situations are – and intelligently so – highly ritual-like.

14. For example, Forester, 1993 shows that a collegial group of land use planners used meeting ritual to sustain patterns of friendship necessary to their institutional style in policy making, which seems, from his account, to be a combination of the central professional unit with some elements of the enclave. Similarly, Schwartzman 1989 shows how the relationships within a highly political agency – which resembles a classical enclave movement organisation – were sustained through meeting ritual. Westrup's analysis of the judgment process in installing an information system in a professionalised context (a combination of the highly constrained, strongly bonded and the more enclave-like weakly constrained, strongly bonded setting, in the terminology of this book) shows that the ritual was necessary to define membership in decision-making groups, to fix roles, to create trust, to reduce uncertainty, and to provide a set of mnemonics to provide materials for sense-making. Peck *et al.* (forthcoming) use the present theory to explain the centrality of the ritual order to the functioning of an enduring joint health and social services meeting in a British county.

15. At this level of analysis, the only remaining possibility is to withdraw altogether from the organisational process (in this case) of policy making, and become a "hermit": Douglas, 1978; Thompson *et al.*, 1990. That position is, by definition, largely irrelevant to the concerns of this book.

16. Thompson 1997a,b,c calls such four-way settlements 'clumsy institutions', in order to contrast them with the institutional commitment to sleekness, simplicity, 'purity' (Douglas, 1966a), 'integrity' and 'authenticity' which the four basic styles constantly call for. Because the four basic institutional styles are

'mechanical solidarities' in Durkheim's (1984 [1893a]) sense, each insisting on 'similarity' or 'purity' to its practice of organisation, such clumsy four-way settlements are the most helpful way in which we can reconstruct the relevance of Durkheim's idea that there is something normatively valuable about 'organic solidarity', or institutional settlements between dissimilar forms of organisation.

10 Ways Forward

17. See note 9 above for the debate on this question of interpretation.
18. It is worth acknowledging, of course, that this senior executive hegemony of the competitive individualists was not easily achieved, and in many ways its current form reflects the reforms of the 1980s and 1990s which were designed to buttress this central ministerial and senior executive hegemony over policy making, and to reduce the influence and even power that was felt to have slipped to other institutional settings in the policy process. This was explicitly argued for in much of the public choice literature of the 1970s and 1980s (e.g. Niskanen, 1971) and its effects have been widely commented upon (e.g. from the left, Campbell and Wilson, 1995, and from the right, Jenkins, 1995).
19. This was Mill's view, for example. In the chapter entitled 'Of the executive in representative government' in the *Considerations on representative government*, he explicitly argues for accountability on the grounds of greater wisdom arising from mixed counsel (consistently with the famous consequentialist argument for freedom of speech in *On liberty*), as well as on traditional grounds of the fear of abuse of power: Mill, 1972 [1861], 331–3. Likewise, Madison's equally famous argument for a system of executive accountability that pitted faction against faction in *Federalist papers* LI, is framed both in terms of the protection of liberty and the goal of securing better judgment in policy making: Madison *et al.*, 1987 [1788], 318–22. Again, Constant argued for ministerial accountability to the legislature on the basis that such accountability was a crucial discipline, not merely for liberty, but for the quality of the process and product of political judgment: Constant, 1988, [1815], 221–7.
20. I recognise, of course, that this is to use the names of Burke and Paine to embody highly polarised positions. Paine's real views, in particular, were much more nuanced than this allows (see Paine, 1988, and Whatmore, 2000), although the statement caricatures Burke much less: see Burke, 1981 [1780]. However, this is the most powerful way to make the point.

11 Conclusion

21. Not only the early pioneers: it remains the confident expectation of artificial intelligence practitioners like Professor Kevin Warwick: see Warwick, 1998. For an argument that artificial intelligence cannot eliminate judgment, even when or if it can model it, see 6, 1999c, 2001a,b.

References

6 P, 1997a, *Holistic government*, Demos, London.

6 P, 1997b, Governing by cultures, in Mulgan GJ, ed, 1997, *Life after politics: new thinking for the twenty first century*, HarperCollins, London, 260–85.

6 P, 1998, *The future of privacy, volume I: private life and public policy*, Demos, London.

6 P, 1999a, Neo-Durkheimian institutional theory, paper for the conference, *Institutional theory in political science*, at Ross Priory, Loch Lomond, 17–18 October, University of Strathclyde.

6 P, 1999b, A constitutional culture for more participation: what would it look like?, in Campbell ID and Lewis ND, eds, 1999, *Promoting participation: law or politics?*, Cavendish, London, 75–100.

6 P, 1999c, *Morals for robots and cyborgs: ethics, society and public policy in the age of autonomous intelligent machines*, Bull Worldwide Information Systems, Brentford, London.

6 P, 2000, The morality of managing risk: paternalism, prevention, precaution and the limits of proceduralism, *Journal of risk research*, 3, 2, 135–65.

6 P, 2001a, Ethics, regulation and the new artificial intelligence, Part I: accountability and power, *Information communication and society*, 4, 2, 199–229.

6 P, 2001b, Ethics, regulation and the new artificial intelligence, Part II: autonomy and liability, *Information, communication and society*, 4, 3, 406–34.

6 P, 2001c, Governing by technique: judgment and the prospects for governance of and with technology, in Organisation for Economic Cooperation and Development, 2001, *Governance in the 21st century*, Organisation for Economic Cooperation and Development, Paris, 67–120.

6 P, 2001d, E-governance: do digital aids make a difference in policy making?, in Prins JEJ, ed, 2001, *Designing e-government: on the crossroads of technological innovation and institutional change*, Kluwer Law International, The Hague, Netherlands, 7–28.

6 P, 2001e, E-governance: Weber's revenge?, in Dowding K, Hughes J and Margetts H, eds, 2001, *Challenges to democracy: ideas, involvement and institutions – the PSA yearbook 2000*, Palgrave, Basingstoke, 220–36.

6 P, 2002a, Can policy making be evidence based?, *Managing community care*, 10, 1, Feb, 3–8, available at http://www.elsc.org.uk/socialcareresource/managecc/feb2002/pages3–8.pdf

6 P, 2002b, Global digital communications and the prospects for transnational regulation, in Held D and McGrew A, eds, 2002, *Governing globalisation: power, authority and global governance*, Polity Press, Cambridge and Blackwell, Oxford, 145–70.

6 P, 2003, in press, What's in a frame? Social organisation, risk perception and the sociology of knowledge, *Journal of risk research*.

6 P forthcoming, 'Organisational memory and capability' in 6 P and Peck E and associates, forthcoming, *Beyond 'delivery': implementation as sense-making*.

6 P and Forder JE, 1996, Can campaigning be evaluated?, *Nonprofit and voluntary sector quarterly*, 25, 2, 225–47.

6 P, Leat D, Seltzer K and Stoker G, 1999, *Governing in the round: strategies for holistic government*, Demos, London.

6 P, Leat D, Seltzer K and Stoker G, 2002, *Toward holistic governance: the new agenda in government reform*, Palgrave, Basingstoke.

Adams J, 1995, *Risk*, UCL Press, London.

Akrich M, 1992, The de-scription of technical objects, in Bijker WE and Law J, eds, 1992, *Shaping technology / building society: studies in sociotechnical change*, Massachusetts Institute of Technology Press, Cambridge, Massachusetts, 205–24.

Allison G and Zelikow P, 1999 [1971], *Essence of decision: explaining the Cuban missile crisis*, 2nd edn, Addison Wesley Longman, New York.

Arendt H, 1954a, The crisis in culture: its social and political significance, in Arendt H, 1954, *Between past and future*, Penguin, Harmondsworth, 197–226.

Arendt H, 1954b, Truth in politics, in Arendt H, 1954, *Between past and future*, Penguin, Harmondsworth, 227–64.

Arendt H, 1963, *Eichmann in Jerusalem*, Penguin, Harmondsworth.

Arendt H, 1992a [1958], *The human condition*, 2nd edn, University of Chicago Press, Chicago, 147–52.

Arendt H, 1992b, *Lectures on Kant's political philosophy*, ed, Beiner R, University of Chicago Press, Chicago.

Arter D, 2000, The model for parliaments in the future? The case of the Finnish Committee for the Future, typescript, Nordic Policy Studies Centre, University of Aberdeen.

Ashby WR, 1947, Principles of the self-organising dynamic system, *Journal of general psychology*, 37.

Bachrach P and Baratz MS, 1962, Two faces of power, *American political science review*, 56, 947–52.

Ballantine J and Cunningham N, 1999, Strategic information systems planning: applying private sector frameworks in UK public healthcare, in Heeks R, ed, 1999, *Reinventing government in the information age: international practice in it-enabled public sector reform*, Routledge, London, 293–311.

Bannon LJ, 1997, A challenge to certain (G)DSS perspectives on the role of decisions, information and technology in organisations, in Humphreys P, Ayestaran S, McCosh A, Mayon-White B, eds, 1997, *Decision support in organisational transformation*, Chapman and Hall, London.

Barker A and Peters BG, eds, 1992, *The politics of expert advice: creating, using and manipulating scientific knowledge for public policy*, Edinburgh University Press, Edinburgh.

Barnard C, 1938, *The functions of the executive*, Harvard University Press, Cambridge, Massachusetts.

Barnard C, 1938b, Mind in everyday affairs, in Barnard C, 1938a, *The functions of the executive*, Harvard University Press, Cambridge, Massachusetts.

Barrett N, 1998, 2010 – the year of the 'digital familiar', in Barrett N, 1998, *Into the third millennium: trends beyond the year 2000*, Bull Information Systems, London, 27–37.

Barry A, Osborne T, and Rose N, eds, 1996, *Foucault and political reason: liberalism, neo-liberalism and rationalities of government*, UCL Press, London.

Bass BM, 1990, *Bass and Stodgill's handbook of leadership*, Free Press, New York.

Beiner R, 1983, *Political judgment*, Methuen, London.

Beiner R, 1992, Interpretive essay: Hannah Arendt on judging, in Arendt H, 1992b, *Lectures on Kant's political philosophy*, ed, Beiner R, University of Chicago Press, Chicago, 89–156.

Beiner R and Nedelsky J, eds, 2001, *Judgment, imagination and politics: themes from Kant and Arendt*, Rowman and Littlefield, Lanham, Maryland.

Bell C, 1997, *Ritual: perspectives and dimensions*, Oxford University Press, New York.

Bellamy C and Taylor JA, 1998, *Governing in the information age*, Open University Press, Buckingham.

Berger PL and Luckmann T, 1966, *The social construction of reality: a treatise in the sociology of knowledge*, Penguin, Harmondsworth.

Berlin I, 1978 [1951], The hedgehog and the fox, in Berlin I, 1978, *Russian thinkers*, Penguin, Harmondsworth, 22–81, orig. pub. as Lev Tolstoy's historical scepticism, in *Oxford Slavonic papers 2*.

Berry FS, Berry JD and Foster SK, 1998, The determinants of success in implementing an expert system in state government, *Public administration review*, 58, 4, 293–305.

Bijker WE, 1987, The social construction of Bakelite: toward a theory of invention, in Bijker WE, Hughes PT and Pinch TJ, eds, 1987, *The social construction of technological systems*, Massachusetts Institute of Technology Press, Cambridge, Massachusetts, 159–87.

Bijker WE, 1995, *Of bicycles, bakelites and bulbs: toward a theory of sociotechnical change*, Massachusetts Institute of Technology Press, Cambridge, Massachusetts.

Bijker WE and Law J, eds, 1994, *Shaping technology / building society*, Massachusetts Institute of Technology Press, Cambridge, Massachusetts.

Borins S, 1998, *Innovating with integrity: how local heroes are transforming American government*, Georgetown University Press, Washington DC.

Bouckaert G, Ormond D and Peters BG, 2000, *A potential governance agenda for Finland: turning 90° in the administration's tasks and functions*, Ministry of Finance, Helsinki, Finland.

Bourdieu P. 1984 [1979], *Distinction: a social critique of the judgment of taste*, Routledge, London.

Bryner G, 1998, *Politics and public morality: the great American welfare reform debate*, WW Norton & Co, New York.

Buckingham Shum S, 1998, Negotiating the construction of organisational memories, in Borghoff UM and Pareschi R, eds, 1998, *Information technology for knowledge management*, Springer-Verlag, Berlin.

Bugler DT and Bretschneider SI, 1993, Technology push or program pull? Interest in new technologies within public organisations, in Bozeman B, ed, 1993, *Public management: the state of the art*, Jossey-Bass, San Francisco, California, 275–93.

Bulmer M, ed, 1987, *Social science research and government*, Cambridge University Press, Cambridge.

Burchell G, Gordon C, and Miller P, eds, 1991, *The Foucault effect: studies in governmentality*, Harvester Wheatsheaf, Hemel Hempstead.

Burke E, 1981 [1780], Speech to the electors of Bristol, in Burke E, 1981, *The writings and speeches of Edmund Burke, Volume 1*, ed, Langford P, 1981, Oxford University Press, Oxford.

Burke WW, 2002, *Organisation change: theory and practice*, Sage, London.

Byrne D, 1998, *Complexity theory and the social sciences: an introduction*, Routledge, London.

Caiden N, 1998, A new generation of budget reforms, in Peters BG and Savoie DJ, eds, 1998, *Taking stock: assessing public sector reforms*, Canadian Centre for Management Development, McGill-Queen's University Press, Montreal, 252–84.

Campbell C and Wilson GK, 1995, *The end of Whitehall: death of a paradigm?*, Blackwell, Oxford.

Carter N, Klein RE and Day P, 1993, *How organisations measure success: the use of performance indicators in government*, Routledge, London.

Castells M, 1996, *The information age: economic, society and culture, volume 1 – the rise of the network society*, Blackwell, Oxford.

Caudle S, 1990, Managing information resources in state government, *Public administration review*, 50, 5, 515–25.

Cavaye ALM, 1998, An exploratory study investigating transnational information systems, *Journal of strategic information systems*, 7, 17–35.

Checkland PB, 1994, Systems theory and management thinking, in Blunden M and Dando M, eds, 1994, *Rethinking public policy-making: questioning assumptions, challenging beliefs – essays in honour of Sir Geoffrey Vickers on his centenary*, Sage, London, 75–91.

Checkland PB, 2001 [1989], Soft systems methodology, in Rosenhead and Mingers J, eds, 2001 [1989], *Rational analysis for a problematic world: problem structuring methods for complexity, uncertainty and conflict*, John Wiley and Sons, Chichester, . 61–90.

Checkland PB, and Scholes J, 1990, *Soft systems methodology in action*, John Wiley and Sons, Chichester.

Cinderby S, Forrester J, Rosen P and Yearley S, 2002, Participation, models and the local governance of air pollution, paper presented at the European Association for the Study of Science and Technology conference, York.

Cloete F, 2001a, Improving good governance with electronic policy management assessment tools, paper presented at the Public Futures 2nd annual conference, 28 Sept, London, available at http://www.emergence.nu/events/wwweprog.html

Cloete F, 2001b, Facilitating negotiations through electronic decision support software, available at http://www.emergence.nu/events/wwweprog.html

Cloete F, 2001c, Improving effective governance outcomes with electronic decision support tools, available at http://www.emergence.nu/events/wwweprog.html

Committee for the Future, 1996, *Models of success, part 1: Southeast and East Asia*, tr. The English Centre, The Finnish Parliament, Helsinki.

Committee for the Future, 1997a, *Models of success, part 2: China*, tr. The English Centre, The Finnish Parliament, Helsinki.

Committee for the Future, 1997b, *Finland-Wisconsin; a comparative study focusing on the job opportunities for Finland*, tr. The English Centre, The Finnish Parliament, Helsinki.

Committee for the Future, 1998, *Skill and fair play: an active and responsible Finland*, tr. The English Centre, The Finnish Parliament, Helsinki.

Committee for the Future, 1999, *Models of success, part 3: Japan*, tr. The English Centre, The Finnish Parliament, Helsinki.

Committee for the Future, 2000a, *Painspots in the future of work*, tr. The English Centre, The Finnish Parliament, Helsinki.

Committee for the Future, 2000b, *New elements in the world of knowledge*, tr. The English Centre, The Finnish Parliament, Helsinki.

Conklin J, 1999, Seven years of industrial strength CSCA in an electric utility, published at http://d3e.open.ac.uk/cscl99/Conklin/Conklin-paper.html

Constant B, 1988 [1815], *Principles of politics applicable to all representative governments*, in Fontana B, ed, 1988, *Constant: political writings*, Cambridge University Press, Cambridge, 170–307.

Coyle DJ and Ellis RJ, eds, 1993, *Politics, policy and culture*, Westview Press, Boulder, Colorado.

Crick B, 1964 [1962], *In defence of politics*, 2nd edn, Penguin, Harmondsworth.

Davies HTO, Nutley SM and Smith PC, eds, 2000, *What works? Evidence-based policy and practice in public services*, Policy Press, Bristol.

Dean M, 1999, *Governmentality: power and rule in modern society*, Sage, London.

DeLeon P, 1999, The stages approach to the policy process: what has it done? where is it going?, in Sabatier PA, ed, 1999, *Theories of the policy process*, Westview, Boulder, Colorado, 1999, 19–32.

Denhardt RB, 1993, *The pursuit of significance: strategies for managerial success in public organisations*, Westview Press, Boulder, Colorado.

Dennis AR, 1996, Information exchange and use in group decision making: you can lead a group to information but you can't make it think, *Management information systems quarterly*, 20, 4, 433–58.

Dennis AR, Valacich JS, Carte TA, Garfield MJ, Haley BJ and Aronson JE, 1997, Research report: the effectiveness of multiple dialogues in electronic brainstorming, *Information systems research*, 8, 2, 203–11.

D'Entrèves MP, 1994, *The political philosophy of Hannah Arendt*, Routledge, London.

De Tocqueville A, 1998 [1835 and 1840], *Democracy in America*, tr Reeve H and rv. Bowen F, abr. Renshaw P, Wordsworth, Ware.

Douglas M, 1966a, *Purity and danger: an analysis of the concepts of pollution and taboo*, Routledge, London.

Douglas M, 1966b, The contempt of ritual, *New society*, 31.3.66, repr. in Douglas, M, 1982b, *In the active voice*, Routledge and Kegan Paul, London, 34–8.

Douglas M, 1970, Natural symbols: explorations in cosmology, Routledge and Kegan Paul, London.

Douglas M, 1972, Self-evidence, reproduced in Douglas M, 1999 [1975], *Implicit meanings: selected essays in anthropology*, 2nd edn, Routledge, London, 252–83.

Douglas M, 1978, Cultural bias, reprinted in Douglas M, 1982b, *In the active voice*, Routledge and Kegan Paul, London, 183–254.

Douglas M, 1980, *Evans-Pritchard: his life, work, writings and ideas*, Harvester Press, Brighton.

Douglas M, 1981, Good taste: review of Pierre Bourdieu, *La distinction*, Times Literary Supplement, 13.2.81, 163–9, repr. in Douglas M, 1982b, *In the active voice*, Routledge and Kegan Paul, London, 125–34.

Douglas M, ed, 1982a, *Essays in the sociology of perception*, Routledge and Kegan Paul, London.

Douglas M, 1982b, *In the active voice*, Routledge and Kegan Paul, London.

Douglas M, 1986, *How institutions think*, Routledge and Kegan Paul, London.

Douglas M, 1992a, *Risk and blame: essays in cultural theory*, Routledge, London.

Douglas M, 1992b, Why do people want goods?, in Hargreaves-Heap S and Ross, A, eds, 1992, *Understanding the enterprise culture: themes in the work of Mary Douglas*, Edinburgh University Press, Edinburgh, 19–31.

Douglas M, 1996, *Thought styles: critical essays on good taste*, Sage, London.

Douglas M and Isherwood B, 1979, *The world of goods: toward an anthropology of consumption*, Routledge, London.

Douglas M and Mars G, 2003, Terrorism: a positive feedback game, *Human relations*, 56, 7, 763–786.

Dowding K, 1995, Model or metaphor? A critical review of the policy network approach, *Political studies*, 43, 1, 136–158.

Dowie J and Elstein A, eds, 1988, *Professional judgment: a reader in clinical decision making*, Cambridge University Press, Cambridge.

Dreyfus HL and Dreyfus S, 1986, *Mind over machine*, MacMillan, Basingstoke.

Durkheim É, 1906, Review of Joseph Hitier, *La doctrine de l'absolutisme*, Paris: Arthur Rousseau, 1903, 228pp, in *L'annee sociologique*, 9, 353–5, tr. by J French, in Nandan Y, ed, 1980, *Emile Durkheim: contributions to* L'annee sociologique, Free Press, New York, 357.

Durkheim É, 1951 [1897a], *Suicide: a study in sociology*, tr. Spaulding JA and Simpson G, Routledge, London.

Durkheim É, 1961, [1925a], *Moral education: a study in the theory and application of the sociology of education*, tr Wilson EK and Schnurer H, Free Press, New York.

Durkheim É, 1982 [1895a], *The rules of sociological method and selected texts on sociology and its method*, tr. Halls WD, MacMillan, Basingstoke.

Durkheim É, 1984 [1893a], *The division of labour in society*, tr. Halls WD, MacMillan, Basingstoke.

Durkheim É, 1995 [1912a], *Elementary forms of religious life*, tr. Fields KE, Free Press, New York.

Durkheim É and Mauss M, 1963 [1903], *Primitive classification*, tr. Needham R, University of Chicago Press, Chicago [Durkheim É and Mauss M, 1903, De quelques formes primitives de classification: contribution à l'etude des représentations collectives, *Année sociologique*, vol. VI, for 1901–2, 1–72.].

Edelman M, 1985 [1964], *The symbolic uses of politics*, University of Illinois Press, Urbana, Illinois.

Edelman M, 1988, *Constructing the political spectacle*, University of Chicago Press, Chicago.

Elgin CZ, 1996, *Considered judgment*, Princeton University Press, Princeton, New Jersey.

Elkin SL and Soltan KE, eds, 1999, *Citizen competence and democratic institutions*, Pennsylvania State University Press, University Park, Pennsylvania.

Ellis RJ and Thompson M, eds, 1997, *Culture matters: essays in honour of Aaron Wildavsky*, Westview Press, Boulder, Colorado.

Eve RA, Horsfall S and Lee ME, eds, 1997, *Chaos, complexity and sociology: myths, models and theories*, Sage, London.

Feldman MS, 1989, *Order without design: information production and policy making*, Stanford University Press, Stanford, California.

Finlay PN and Forghani M, 1998, A classification of success factors for decision support systems, *Journal of strategic information systems*, 7, 53–70.

Fischer F, 2003, *Reframing public policy: discursive politics and deliberative practices*, Oxford University Press, Oxford.

Fleck L, 1979 [1935], *The genesis and development of a scientific fact*, tr. Bradley F and Trenn TJ, ed, Trenn TJ and Merton RK, University of Chicago Press, Chicago.

Fleischacker S, 1999, *A third concept of liberty: judgment and freedom in Kant and Adam Smith*, Princeton University Press, Princeton, New Jersey.

Forester J, 1993. Learning from practice stories: the priority of practical judgment, in Fischer F and Forester J, eds, 1993, *The argumentative turn in policy analysis and planning*, UCL Press, London, 186–209.

Forester J, 1999, Beyond dialogue to transformative learning: how deliberative rituals encourage political judgment in community planning processes, in Forester J, 1999, *The deliberative practitioner: encouraging participatory planning processes*, Massachusetts Institute of Technology Press, Cambridge, Massachusetts, 115–53.

Forrester J, Potts L, Cinderby S and Rosen P, 2002, Creating 'science's new social contract'?, paper presented at the European Association for the Study of Science and Technology conference, York.

Foucault M, 1991, Governmentality, in Burchell G, Gordon C and Miller P, eds, 1991, *The Foucault effect: studies in governmentality*, Harvester Wheatsheaf, Hemel Hemstead, 87–104.

Fountain JE, 2001, *Building the virtual state: information technology and institutional change*, Brookings Institution, Washington DC.

Fox N, and Roberts C, 1999, GPs in cyberspace: the sociology of a 'virtual community', *The sociological review*, 4, 643–71.

Frederickson HG and Smith KB, 2003, *The public administration theory primer*, Westview Press, Boulder, Colorado.

Frissen PHA, 1999, *Politics, governance and technology: a postmodern narrative on the virtual state*, Edward Elgar, Cheltenham.

Frøkjaer E, 1989, Controversial issues of expert systems in public administration, in Snellen ITHM, van de Donk WBHJ and Baquiast J-P, eds, 1989, *Expert systems in public administration*, Elsevier, Amsterdam, 35–54.

Frost P and Robinson S, 1999, The toxic handler: organisational hero and casualty, *Harvard business review*, Jul–Aug, 96–106.

Gamble A, 2000, *Politics and fate*, Polity Press, Cambridge.

Gamson WA, 1992, *Talking politics*, Cambridge University Press, Cambridge.

Gardner H, 1997, *Leading minds: the anatomy of leadership*, HarperCollins, London.

Giuliani R, 2002, *Leadership*, Miramax, New York.

Goffman E, 1967, *Interaction ritual: essays on face-to-face behaviour*, Pantheon, New York.

Goffman E, 1974, *Frame analysis: an essay on the organisation of experience*, Harper and Row, New York.

Goodman N, 1978, *Ways of worldmaking*, Hackett Publishing, Indianapolis, Indiana.

Greenberger M, Crenson MA and Crissey BL, 1976, *Models in the policy process: decision making in the computer era*, Russell Sage Foundation, New York.

Grønbjerg KA, 1993, *Understanding nonprofit funding: managing revenues in social services and community development organisations*, Jossey-Bass, San Francisco, California.

Gross JL and Rayner S, 1985, *Measuring culture: a paradigm for the analysis of social organisation*, Columbia University Press, New York.

Gulick L and Urwick L, eds, 1937, *Papers in the science of administration*, Institute of Public Administration, New York.

Hacking I, 1990, *The taming of chance*, Cambridge University Press, Cambridge.

Hague BN and Loader BD, eds, 1999, *Digital democracy: discourse and decision making in the information age*, Routledge, London.

Hammer M, and Champy J, 1995, *Re-engineering the corporation: a manifesto for business revolution*, Nicholas Brearley Publishing, London.

Hammond K, 1996, *Human judgment and social policy: irreducible uncertainty, inevitable error and unavoidable injustice*, Oxford University Press, New York.

Hargreaves Heap S and Ross A, eds, 1992, *Understanding the enterprise culture: themes in the work of Mary Douglas*, Edinburgh University Press, Edinburgh.

Hasan H and Hasan S, 1997, Computer-based performance information for executives in local government, *Australian journal of public administration*, 56, 3, 24–9.

Hastie R and Dawes RM, 2001, *Rational choice in an uncertain world: an introduction to judgment and decision making*, Sage, London.

Hayes N and Walsham G, 2000, Competing interpretations of computer-supported cooperative work in organisational contexts, *Organisation*, 7, 1, 49–67.

Heinrich C, 1999, Do government bureaucrats make effective use of performance management information?, *Journal of public administration research and theory*, 9, 3, 363–94.

Hogwood BA, Judge D, and McVicar M, 1998, Agencies and accountability, paper presented to the Whitehall Programme conference, University of Birmingham, 17–19 December 1998.

Hogwood BW and Gunn LA, 1984, *Policy analysis for the real world*, Oxford University Press, Oxford.

Holtham CW, 2000, Interim electronic government may be major barrier to full electronic government, *Public finance and accountancy*, May.

Hood CC, 1983, *The tools of government*, MacMillan, Basingstoke.

Hood CC, 1998, *The art of the state: culture, rhetoric and public management*, Oxford University Press, Oxford.

Hunter P, 1999, Spies on your hard drive: software agents, *Computer weekly*, 30.9.99, 32.

Immergut EM, 1992, *Health politics: interests and institutions in western Europe*, Cambridge University Press, Cambridge.

Innvaer S, Vist G, Trommald M, and Oxman A, 2002, Health policy-makers' perceptions of their use of evidence: a systematic review, *Journal of health services research and policy*, 7, 4, 239–44.

Jackson PM, 1992, Economic policy, in Marsh D and Rhodes RAW, eds, 1992, *Implementing Thatcherite policies: audit of an era*, Open University Press, Buckingham, 11–31.

Janis IR, 1982, *Victims of groupthink*, 2nd edn, Houghton Mifflin, Boston.

Jasanoff S, Markle GE, Petersen JC and Pinch T, eds, 1994, *Handbook of science and technology studies*, Sage, Newbury Park, California.

Jenkins S, 1995, *Accountable to none: the Tory nationalisation of Britain*, Penguin, Harmondsworth.

Jepperson RL, 1991, Institutions, institutional effects and institutionalism, in Powell WW and DiMaggio PJ, eds, 1991, *The new institutionalism in organisational analysis*, University of Chicago Press, Chicago, Illinois, 143–63.

John P, 1998, *Analysing public policy*, Pinter, London.

Jones MR, 1994, Information technology for group decision support: beyond GDSS, *Journal of organisational computing*, 4, 1, 23–40.

Kahneman D, Slovic P and Tversky A, eds 1982, *Judgment under uncertainty: heuristics and biases*, Cambridge University Press, Cambridge.

Kahneman D and Tversky A, 2000, *Choices, values and frames*, Cambridge University Press, Cambridge.

Kamarck EC and Nye JS, Jr, 2002, *Governance.com: democracy in the information age*, Brookings Institution, Washington DC.

Kant I, 1951, *Critique of judgment*, tr. Bernard JH, Hafner, New York.

Kant I, 1983 [1784], An answer to the question, 'What is enlightenment?', in Kant I, 1983, *Perpetual peace and other essays*, ed, Humphrey T, Hackett Publishing, Indianapolis, Indiana.

Kant I, 1987, *Fundamental principles of the metaphysic of morals*, tr. Abbott TK, Prometheus Books, Buffalo, New York.

Karagiannis D, Radermacher FJ, Teufel B and Wynne BE, 1994, Towards CSCW: metalevel environments for enhanced group and organisation effectiveness, *Journal of organisational computing*, 4, 4, 367–92.

Kateb G, 2001, The judgment of Arendt, in Beiner R and Nedelsky J, eds, 2001, *Judgment, imagination and politics: themes from Kant and Arendt*, Rowman and Littlefield, Lanham, Maryland.

Kendall J, 2003, *The voluntary sector: comparative policy and impacts in the UK*, Routledge Curzon, London.

Kingdon JW, 1995 [1984], *Agendas, alternatives and public policies*, 2nd edn, HarperCollins, New York.

Klein M, Roux D and Villedieu T, 1991, Decision support for municipality financial planning in France: recent progress with 'SIAD Mairie', a knowledge based DSS, in Traunmüller R, ed, 1991, *Governmental and municipal information systems II*, proceedings of the 2nd IFIP TC8/WG8.5 Working conference on governmental and municipal information systems, Balatonfüred, Hungary, North-Holland, Amsterdam, 117–30.

Kling R, 1997, Multivalent social relationships in computer supported workplaces, in Kiesler S, ed, 1997, *Research milestones on the information highway*, Lawrence Erlbaum, Hillsdale, New Jersey.

Knight J, 1992, *Institutions and social conflict*, Cambridge University Press, Cambridge.

Knoke D, 1990, *Political networks: the structural perspective*, Cambridge University Press, Cambridge.

Kooiman J, 2003, *Governing as governance*, Sage, London.

Kraemer K and Dutton WJ, 1982, The automation of bias, in Danziger JN, Dutton WH, Kling R and Kraemer KL, 1982, *Computers and politics: high technology in American local governments*, Columbia University Press, New York, 136–69.

Kraemer KL and Dedrick J, 1997, Computing and public organisations, *Journal of public administration research and theory*, 7, 1, 89–112.

Kroon N, 1997, European information systems: shifting boundaries (title translated from Dutch), Ph.D thesis, Erasmus University, Rotterdam.

Kuhn T, 1970 [1962], *The structure of scientific revolutions*, University of Chicago Press, Chicago.

Landell-Mills P, 2003, E-government: new tools for policy accountability, *Oxford Policy Institute issues notes*, 4, February, 3–5.

Landsbergen D, Coursey DH, Loveless S and Shangraw RF Jr, 1997, Decision quality, confidence and commitment with expert systems: an experimental study, *Journal of public administration research and theory*, 7, 1, 131–57.

Lasswell H, 1956, *The decision process*, University of Maryland Press, College Park, Maryland.

Laumann EO and Knoke D, 1987, *The organisational state*, University of Wisconsin Press, Madison, Wisconsin.

Lee AS, 1994, Electronic mail as a medium for rich communication: an empirical investigation using hermeneutic interpretation, *Management information systems quarterly*, 18, 2, 143–58.

Leeuw FL, Rist RC and Sonnichsen RC, eds, 1994, *Can governments learn? Comparative perspectives on evaluation and organisational learning*, Transaction Publishers, New Brunswick, New Jersey.

Leith P, 1986, Fundamental errors in legal logic programming, *The computer journal*, 29, 6, 545–52.

Lenk K, Traunmüller R and Wimmer M, 2002, The significance of law and knowledge for electronic government, in Grönlund Å, ed, 2002, *Electronic government: design, applications and management*, Idea Group, Hershey.

Lipsky M, 1980, *Street level bureaucracy: dilemmas of the individual in public service*, Russell Sage Foundation, New York.

Lockwood D, 1992, *Solidarity and schism: the problem of 'disorder' in Durkheimian and Marxist sociology*, Clarendon Press of the Oxford University Press, Oxford.

Louis M, 1980, Surprise and sense-making: what newcomers experience in entering unfamiliar organisational settings, *Administrative science quarterly*, 5, 226–51.

Mackenzie D and Wajcman J, eds, 1985, *The social shaping of technology: how the refrigerator got its hum*, Open University Press, Buckingham.

Madison J, Hamilton A and Jay J, 1987 [1788], *The federalist papers*, Penguin, Harmondsworth.

Majone G, 1989, *Evidence, argument and persuasion in the policy process*, Yale University press, New Haven.

Malloy J, 2003, *To better serve Canadians: how technology is changing the relationship between Members of Parliament and public servants*, Institute of Public Administration of Canada, Toronto.

Mannheim K, 1985 [1936], *Ideology and utopia: an introduction to the sociology of knowledge*, Harcourt Brace Jovanich, San Diego, California.

March JG, 1994, *A primer on decision making: how decisions happen*, Free Press, New York.

March JG, 1999a, *The pursuit of organisational intelligence*, Blackwell, Oxford.

March JG, 1999b [1996], Understanding how decisions happen in organisations, in March JG, 1999, *The pursuit of organisational intelligence*, Blackwell, Oxford, 13–38, originally published as March 1996, A preface to understanding how decisions happen in organisations, in Shapira Z, ed, 1996, *Organisational decision making*, Cambridge University Press, New York.

March JG, 1999c [1991], Exploration and exploitation in organisational learning, *Organisation science*, 2, 71–87, repr in March JG, 1999, *The pursuit of organisational intelligence*, Blackwell, Oxford, 114–136.

March JG and Levinthal DA, 1999 [1993], The myopia of learning, *Strategic management learning*, 14, 95–112., repr. in March JG, 1999, *The pursuit of organisational intelligence*, Blackwell, Oxford.

March JG, with Lounamaa PH, 1999d [1987], Adaptive coordination of a learning team, *Management science*, 33, 107–23, repr in March JG, 1999, *The pursuit of organisational intelligence*, Blackwell, Oxford, 156–78.

March JG and Olsen J-P, 1975, The uncertainty of the past: organisational learning under ambiguity, *European journal of political research*, 3, 147–71, repr in March JG, 1988, *Decisions and organisations*, Blackwell, Oxford, 335–58.

March JG and Olsen J-P, 1976, *Ambiguity and choice in organisations*, Universitetsforlaget, Bergen.

March JG and Olsen J-P, 1989, *Rediscovering institutions: the organisational basis of politics*, Free Press, New York.

March JG and Olsen J-P, 1995, *Democratic governance*, Free Press, New York.

March JG and Simon HA, with Guetzkow H, 1993 [1958], *Organisations*, 2nd edn, Blackwell, Oxford.

Marcus GE, Neumann WR and Mackuen M, eds, 2000, *Affective intelligence and political judgment*, University of Chicago Press, Chicago.

Margetts H and Dunleavy P, 1995, Public services on the world markets, in *Missionary government, Demos collection*, 7, 30–2.

Mars G, 1982, *Cheats at work: an anthropology of workplace crime*, Allen and Unwin, London.

Marsh D, ed, 1998, *Comparing policy networks*, Open University Press, Buckingham.

Marsh D and Rhodes RAW, eds, 1992, *Policy networks in British government*, Oxford University Press, Oxford.

Massetti B, 1998, An empirical examination of the value of creativity support systems on idea generation, *Management information systems quarterly*, 20, 1, 83–98.

Mauss M, 1979 [1935], Body techniques, in Mauss M, 1979, *Sociology and psychology:* essays, tr Brewster B, Routledge and Kegan Paul, London, 95–123, orig pub as Les techniques dur corps, *Journal de psychologie normale et pathologique*, 35, 271–93.

McNurlin BC and Sprague RH, Jr, 1998 [1993], *Information systems management in practice*, 4th edn, Prentice-Hall, Englewood Cliffs, New Jersey.

Merton RK, 1936, The unintended consequences of purposive social action, *American sociological review*, 1, 894–904, excerpted in Sztompka P, ed, 1996, *Robert K Merton on social structure and science*, University of Chicago Press, Chicago, Illinois, 173–82.

Merton RK, 1945, Sociology of knowledge, in Gurvitch G and Moore WE, eds, 1945, *Twentieth century sociology*, Philosophical Library, New York, excerpted in Sztoupkap ed, 1996, *Robert K Merton on social structure and science*, University of Chicago Press, Chicago, 205–22.

Mill JS, 1972 [1861], Considerations on representative government, in Mill JS, *Utilitarianism, on liberty and representative government*, Acton HB, ed, JM Dent & Sons, London.

Milner E, 2000, *Managing information and knowledge in the public sector*, Routledge, London.

Mosier JN and Tammaro SG, 1994, Video teleconference use among geographically dispersed workgroups: a field investigation of usage patterns and user preferences, *Journal of organisational computing*, 4, 4, 343–65.

Mouritzen PE and Svara J, 2002, *Leadership at the apex: politicians and administrators in western local governments*, University of Pittsburgh Press, Pittsburgh, Pennsylvania.

Nalbandian J, 1995, Politics, administration and the city manager, in Frederickson HG, ed, 1995, *Ideal and practice in council–manager government*, International City / Country Association, Washington DC, 84–9.

Nalbandian J, 1999, Council–staff partnership, *Western city*, September, available at http://www.ku.edu/~nalband/consult/draft.html

National Audit Office (NAO), 1999, *Government on the web*, HC 87, Stationery Office, London.

Nedovic-Budic Z and Godschalk DR, 1996, Human factors in the adoption of geographic information systems: a local government case study, *Public administration review*, 56, 6, 554–67.

Newman and Sabherwal R, 1996, Determinants of commitment to information systems development: a longitudinal investigation, *Management information systems quarterly*, 20, 1, 23–54.

Nidumolu SR, Goodman SE, Vogel DR, Dankowitz AK, 1996, Information technology for local administration support: the Governorates Project in Egypt, *Management information systems*, June, 20, 2, 197–225.

Niederman F, Biese CM and Beranek PM, 1996, Issues and concerns about computer supported meetings, *Management information systems quarterly*, 20, 1, 1–22.

Niskanen W, 1971, *Bureaucracy and representative government*, Aldine-Atherton, Chicago, Illinois.

North DC, 1990, *Institutions, institutional change and economic performance*, Cambridge University Press, Cambridge.

Nunamaker JF Jr, Briggs RO, Mittleman DD, Vogel DR, and Balthazard PA, 1996, Lessons from a dozen years of group support systems research: a discussion of lab and field findings, Centre for the Management of Information, MIS Department, Karl Eller Graduate School of Management, University of Arizona, Tucson, Arizona; available in the Ventana 'knowledge library' at <http://www.ventana.com>

O'Connell PE, 2001, Using performance data for accountability: the New York City Police Department's CompStat model of police management, in Abramson MA and Kamensky JM, eds, 2001, *Managing for results 2002*, Rowman and Littlefield, Lanham, Maryland, 179 224.

Olson DR and Torrance N, eds, 1996, *Modes of thought: explorations in culture and cognition*, Cambridge University Press, Cambridge.

Osborne D and Gaebler T, 1991, *Reinventing government: how the entrepreneurial spirit is transforming the public sector*, Plume (Penguin), New York.

Overman ES and Loraine DT, 1994, Information for control: another management proverb?, *Public administration review*, 193–6.

Paine T, 1988, *'Common Sense', 'the Rights of Man' and other essential writings of Thomas Paine*, Plume Books, New York.

Pawson R, 2002, Evidence and policy and naming and shaming, *Policy studies*, 23, 3, 4, 211–30.

Peck E, Gulliver P and 6 P, forthcoming, Meetings as rituals, in 6 P and Peck E and associates, *Beyond 'delivery': policy implementation as sense-making*.

Performance and Innovation Unit, 2000a, *Adding it up: improving analysis and modelling in central government*, Cabinet Office, London.

Performance and Innovation Unit, 2000b, *Wiring it up: Whitehall's management of cross-cutting policies and services*, Performance and Innovation Unit, Cabinet Office, London.

Performance and Innovation Unit, 2000c, *E-gov: electronic government services for the twenty first century*, Performance and Innovation Unit, Cabinet Office, London.

Peters BG, 1999, *Institutional theory in political science: the 'new institutionalism'*, Pinter, London.

Peters BG and Barker A, eds, 1993, *Advising west European governments: enquiries, expertise and public policy*, Edinburgh University Press, Edinburgh.

Peters BG and van Nispen FKM, eds, 1998, *Public policy instruments: evaluating the tools of public administration*, Edward Elgar, Cheltenham.

Phillips LD, 1988, People centred group decision support, in Doukidis G, Land F and Miller G, eds, 1988, *Knowledge based management support systems*, Ellis Horwood, Chichester.

Phillips LD, 1989, *Decision conferences: description, analysis and implications for group decision support*, Decision Analysis Unit Technical Report 89-2, Decision Analysis Unit, London School of Economics and Political Science.

Pierre J and Peters BG, 2000, *Governance, politics and the state*, MacMillan, Basingstoke.

Polanyi K, 1944, *The great transformation: the economic and political origins of our time*, Beacon Press, Boston, Massachusetts.

Polanyi M, 1958, *Personal knowledge: towards a post-critical philosophy*, Routledge and Kegan Paul, London.

Porter DO and Hjern B, 1981, Implementation structures: a new unit of analysis, *Organisation studies*, 2, 211–27, reprinted in abridged form in Hill M, ed, 1993, *The policy process: a reader*, Harvester Wheatsheaf, Hemel Hempstead, 248–65.

Portman ECP, 1988, The Alvey DHSS Large Demonstrator Project, in Duffin P, ed, *Knowledge-based systems: applications in administrative government*, Ellis Horwood, Chichester and CCTA, London, 76–88.

Power M, 1997, *The audit society: rituals of verification*, Oxford University Press, Oxford.

Pratchett L, 1999, New technologies and the modernisation of local government, *Public administration*, 77, 4, 731–50.

Prinz W and Syri A, 1997, Two complementary tools for the co-operation in a ministerial environment, *Journal of universal computer science*, 3, 8, 843–64.

Proehl RA, 2001, *Organisational change in the human services*, Sage, London.

Raab C, 1997, Privacy, information and democracy, in Loader BD, ed, 1997, *The governance of cyberspace: politics, technology and global restructuring*, Routledge, London, 155–74.

Ranerup A, 1999, Internet-enabled applications for local government democratisation: contradictions of the Swedish experience, in Heeks R, ed, 1999, *Reinventing government in the information age: international practice in IT-enabled public sector reform*, Routledge, London, 177–93.

Rayner S, 1982, The perceptions of time and space in egalitarian sects: a millenarian cosmology, in Douglas M, ed, 1982, *Essays in the sociology of perception*, Routledge and Kegan Paul, London.

Rayner S, 1988, The rules that keep us equal: complexity and costs of egalitarian organisation, in Flanagan JG and Rayner S, eds, 1988, *Rules, decisions and inequality in egalitarian societies*, Averbury, Aldershot, 20–42.

Renshon SA and Larson DW, eds, 2002, *Good judgment in foreign policy: theory and application*, Rowman and Littlefield, Lanham, Maryland.

Reschenthaler GB, and Thompson F, 1996 The information revolution and the new public management, *Journal of public administration research and theory*, 6, 1, 125–43.

Rohrbaugh J, 2000, The use of system dynamics on decision conferencing: implementing welfare reform in New York State, in Garson GD, ed, 2000, *Handbook of public information systems*, Marcel Dekker, New York, 521–33.

Romm (Livermore) CT, 1999, *Virtual politicking: playing politics in electronically linked organisations*, Hampton Press, Cresskill, New Jersey.

Rose N, 1999, *Powers of freedom: reframing political thought*, Cambridge University Press, Cambridge.

Rosenhead J and Mingers J, ed, 2001 [1989], *Rational analysis for a problematic world revisited: problem structuring methods for complexity, uncertainty and conflict*, 2nd edn, Chichester: John Wiley & Son.

Sabatier PA, 1991, Toward better theories of the policy process, *PS: Political science and politics*, 24, June, 147–56.

Sabatier PA 1999, *Theories of the policy process*, Westview Press, Boulder, Colorado.

Sabatier PA and Jenkins-Smith H, eds, 1993, *Policy change and learning: an advocacy coalition approach*, Westview Press, Boulder, Colorado.

Sabatier PA and Jenkins-Smith H, 1999, The advocacy coalition framework: an assessment, in Sabatier PA 1999, *Theories of the policy process*, Westview Press, Boulder, Colorado.

Salamon LM with Lund MS, ed, 1989, *Beyond privatisation: the tools of government action*, Urban Institute, Washington DC.

Salamon LM, ed, 2002, *The tools of government*, Oxford University Press, New York.

Schutz A, 1967, *The phenomenology of the social world*, Heinemann, London.

Schwabe G, 2000, E-councils: systems, experiences, perspectives, paper presented at the 11th International Workshop on Database and Expert Systems Applications (DEXA'00), September 6–8, 2000, Greenwich, London.

Schwabe G and Krcmar H, 2000, Electronic meeting support for councils, *AI and society*, 14, S. 48–70.

Schwarz M and Thompson M, 1990, *Divided we stand: redefining politics, technology and social choice*, Harvester Wheatsheaf, London.

Schwarzman HB, 1989, *The meeting: gatherings in organisations and communities*, Plenum Press, New York.

Scott WR, 1992, *Organisations: rational, natural and open systems*, 3rd edn, Prentice-Hall, Englewood Cliffs, New Jersey.

Senge PM, 1990, *The fifth discipline: the art and practice of the learning organisation*, Century Business [Random House], London.

Shapiro D, Hughes J, Harper R, Ackroyd S and Soothill, K, 1991, Policing information systems: the social context of success and failure in introducing information systems in the police service, in Traunmüller R, ed, 1991, *Governmental and municipal information systems II*, proceedings of the 2nd IFIP TC8/WG8.5 Working conference on governmental and municipal information systems, Balatonfüred, Hungary, North-Holland, Amsterdam, 183–97.

Simon HA, 1977 [1960], *The new science of management decision*, 3rd edn, Prentice-Hall, Englewood Cliffs, New Jersey.

Simon HA, 1992a [1964], Information processing in computer and man, in Simon HA with Egidi M, Marris R and Viale R, 1992, *Economics, bounded rationality and the cognitive revolution*, Edward Elgar, Aldershot, 76–101.

Simon HA, 1992b [1966], Thinking by computers, in Simon HA with Egidi M, Marris R and Viale R, 1992, *Economics, bounded rationality and the cognitive revolution*, Edward Elgar, Aldershot, 55–75.

Simon HA, 1996 [1969], *The sciences of the artificial*, 3rd edn, Massachusetts Institute of Technology Press, Cambridge, Massachusetts.

Simon HA, 1997 [1945], *Administrative behaviour: a study of decision-making processes in administrative organisations*, 4th edn, [1st edn 1945], Free Press, New York.

Slovic P, 2000, *The perception of risk*, Earthscan, London.

Snellen IThM, 1989, Knowledge-based systems in public administration: evolving practices and norms, in Snellen IThM, van de Donk WBHJ and Baquiast J-P, eds, 1989, *Expert systems in public administration*, Elsevier, Amsterdam, 3–22.

Snellen IThM, 1998, Street level bureaucracy in an information age, in Snellen IThM and van de Donk WBJH, eds, 1998, *Public administration in an information age: a handbook*, IOS Press, Amsterdam, 497–508.

Sprague RH Jr and Watson H, 1996, *Decision support for management*, Prentice-Hall, Upper Saddle River, New Jersey.

Steinberger P, 1993, *The concept of political judgment*, University of Chicago Press, Chicago.

Stevens JM and McGowan RP, 1985, *Information systems for public management*, Praeger, New York.

Stokey E and Zeckhauser R, 1978, *A primer for policy analysis*, WW Norton & Co, New York.

Stone D, 2002 [1988], *Policy paradox: the art of political decision making*, WW Norton & Co, New York.

Strategic Policy Making Team, 1999, *Professional policy making for the twenty first century*, Strategic Policy Making Team, Cabinet Office, London.

Sullivan H and Skelcher C, 2002, *Working across boundaries: collaboration in public services*, Palgrave, Basingstoke.

Tapscott D, 1997a, The digital media and the reinvention of government, *Canadian public administration*, 40, 328–45.

Tapscott D, 1997b, *The digital economy: promise and peril in the age of networked intelligence*, McGraw-Hill, New York.

Thompson M, 1996, *Inherent relationality an anti-dualist approach to institutions*, Los Senteret (Norwegian Research Centre in Organisation and Management), Bergen.

Thompson M, 1997a, Rewriting the precepts of policy analysis, in Ellis RJ and Thompson M, eds, 1997, *Culture matters: essays in honour of Aaron Wildavsky*, Westview Press, Boulder, Colorado, 20–216.

Thompson M, 1997b, Cultural theory and technology assessment, in Fischer F and Hajer M, eds, 1997, *Living with nature: environmental discourse as cultural politics*, Oxford University Press, Oxford.

Thompson M, 1997c, Cultural theory and integrated assessment, *Environmetal modelling and assessment*, 2, 139–150.

Thompson M, Ellis RJ and Wildavsky A, 1990, *Cultural theory*, Westview Press, Boulder, Colorado.

Thompson M, Grendstad G and Selle P, eds, 1999, *Cultural theory as political science*, Routledge, London.

Thompson M and Wildavsky A, 1986, A cultural theory of information bias in organisations, *Journal of management studies*, 23, 3, 273–86.

Tiihonnen P, 1997, What will the state do in the 21st century? in Committee for the Future, 1997a, 57–71.

Tiihonnen P, 2000, The internet and politics, paper presented at the 2nd Parliamentary Conference on Information and Communications Technologies, Athens, 3.2.2000.

True JL, Jones BD and Baumgartner FR, 1999, Punctuated equilibrium theory: explaining stability and change in American policy making, in Sabatier PA, ed, 1999, *Theories of the policy process*, Westview Press, Boulder, Colorado, 97–116.

Tversky A and Kahneman D, 1981, The framing of decisions and the psychology of choice, Science, 211, 453–458, repr in Elster J, ed, 1986, *Rational choice*, Oxford University Press, Oxford, 123–141.

van de Donk WBHJ, 1998, Beyond incrementalism? Redistributive policy making, information systems and the revival of synopticism, in Snellen IThM and van de Donk, WBHJ, eds, 1998, *Public administration in an information age: a handbook*, IOS Press, Amsterdam, 381–404.

Van Gunsteren HR, 1976, *The quest for control: a critique of the rational-central rule approach in public affairs*, John Wiley and Sons, London.

Vickers Sir G, 1995 [1965], *The art of judgment: a study in policy making*, centenary edn, Sage, London.

Vogel D, 1998, The globalisation of pharmaceutical regulation, *Governance*, 11, 1, 1–22.

Ward M, 1999, *Virtual organisms: the startling world of artificial life*, MacMillan, Basingstoke.

Warwick K, 1998, *In the mind of the machine: the breakthrough in artificial intelligence*, Arrow, London.

Watson RT, Alexander MB, Pollard CE, and Bostrom RP, 1994, Perceptions of facilitators of a keypad-based group support system, *Journal of organisational computing*, 4, 2, 103–5.

Weatherall A and Nunamaker JF Jr, 1995, *Introduction to electronic meetings*, Electronic Meetings Services Ltd, Chandlers Ford.

Weaver RK and Rockman BA, 1993, *Do institutions matter? Government capabilities in the United States and abroad*, Brookings Institution, Washington DC.

Weber M, 1958 [1946], Bureaucracy, from Weber M, 1946, *Economy and society*, in Gerth HH and Mills CW, eds, 1958 [1946] *From Max Weber: essays in sociology*, Galaxy, New York and Oxford University Press, New York, 196–245.

Weber M, 1976, *The Protestant ethic and the spirit of capitalism*, Allen and Unwin, London.

Weick KE, 1995, *Sensemaking in organisations*, Sage, London.

Weick KE, 2001, *Making sense of the organisation*, Blackwell, Oxford.

Weick KE and Meader D, 1993, Sense making support systems, in Jessup LM and Valecich, JS, eds, 1993, *Group support systems: new perspectives*, MacMillan, New York, 230–52.

Weiss JA, 1979, The many meanings of research utilisation, *Public administration review*, 39, 5, 426–31.

Weiss JA, 1998, Have we learned anything new about the use of the evaluation?, *American journal of evaluation*, 19, 1, 21–33.

Weizenbaum J, 1984 [1976], *Computer power and human reason: from judgment to calculation*, Penguin, Harmondsworth.

Westrup C, 1996, The play of information systems development: drama and ritual in the development of a nursing information systems, *Information technology and people*, 9, 2, 24–42.

Whatmore R, 2000, in Collini S, Whatmore R and Young B, eds, 2000, *Economy, polity and society: British intellectual history, 1750–1950*, Cambridge University Press, Cambridge, 135–57.

Whitby B, 1996, *Reflections on artificial intelligence: the legal, moral and ethical dimensions*, Intellect Books, Exeter.

White M, Howells A, Kibby P and Abell A, 1998, *Intranet management: a guide to best practice*, TFPL, London.

Wiener N, 1948, *Cybernetics: the emerging science at the edge of order and chaos*, Simon and Schuster, New York.

Wildavsky A, 1980, *Speaking truth to power: the art and craft of policy analysis*, Little Brown, Boston, Massachusetts.

Wildavsky A, 1988, *Searching for safety*, Transaction Books, New Brunswick, New Jersey.

Wilhelm AG, 2000, *Democracy in the digital age: challenges to political life in cyberspace*, Rotuledge, London.

Wilson F, 1999, Cultural control within the virtual organisation, *The sociological review*, 4, 672–94.

Winner L, 1977, *Autonomous technology: technics-out-of-control as a theme in political thought*, Massachusetts Institute of Technology Press, Cambridge, Massachusetts.

Winner L, 1980, Do artefacts have politics?, *Daedalus*, 109, 121–36.

Wood M, Pervan G and Schmidenberg O, 1998, The impact of computer modelling on the development of commitment in the application of decision conferencing to strategic issues: preliminary results from a field study, paper presented at the ACIS 1998 Conference.

Yankelovich D, 1991, *Coming to public judgment: making democracy work in a complex world*, Syracuse University Press, Syracuse, New York.

Yanow D, 1996, *How does a policy mean? Interpreting policy and organisational actions*, Georgetown University Press, Washington DC.

Yar M, 2000, From actor to spectator: Hannah Arendt's 'two theories' of political judgment, *Philosophy and social criticism*, 26, 2, 1–27.

Ytterstad P and Watson RT, 1996, Teledemocracy: using information technology to enhance political work, *Management information systems quarterly*, Sept, 20, 3, 347, full details at http://misq.org/discovery.home.html

Zerubavel E, 1997, *Social mindscapes: an invitation to cognitive sociology*, Harvard University Press, Cambridge, Massachusetts.

Zigurs I and Kozar KA, 1994, An exploratory study of roles in computer supported groups, *Management information systems quarterly*, 18, 3, 277–98.

Zuurmond A, 1998, From bureaucracy to infocracy: are democratic institutions lagging behind?, in Snellen I and van de Donk W, eds, 1998, *Public administration in an information age: a handbook*, IOS Press, Amsterdam, 259–72.

Index

GPSR Compliance
The European Union's (EU) General Product Safety Regulation (GPSR) is a set
of rules that requires consumer products to be safe and our obligations to
ensure this.

If you have any concerns about our products, you can contact us on

ProductSafety@springernature.com

In case Publisher is established outside the EU, the EU authorized
representative is:

Springer Nature Customer Service Center GmbH
Europaplatz 3
69115 Heidelberg, Germany